Climate Change Law in China in Global Context

In *Climate Change Law in China in Global Context*, seven climate change law scholars explain how the country's legal system is gradually being mobilized to support the reduction of greenhouse gas emissions in China and achieve adaptation to climate change.

There has been little English scholarship on the legal regime for climate change in China. This volume addresses this gap in the literature and focuses on recent attempts by the country to build defences against the impacts of climate change and to meet the country's international obligations on mitigation. The authors are not only interested in China's laws on paper; rather, the book explains how these laws are implemented and integrated in practice and sheds light on China's current laws, laws in preparation, the changing standing of law relative to policy, and the further reforms that will be necessary in response to the 2015 Paris Agreement on Climate Change.

This comprehensive and critical account of the Chinese legal system's response to the pressures of climate change will be an important resource for scholars of international law, environmental law, and Chinese law.

Xiangbai He is Associate Professor at Guanghua Law School, Zhejiang University, China.

Hao Zhang is Assistant Professor at the Faculty of Law, The Chinese University of Hong Kong (CUHK Law).

Alexander Zahar is Honorary Professor at Macquarie Law School, Sydney, Australia, and Professor of Law at Southwest University of Political Science and Law, Chongqing, China.

Routledge Advances in Climate Change Research

Communication Strategies for Engaging Climate Skeptics
Religion and the Environment
Emma Frances Bloomfield

Regenerative Urban Development, Climate Change and the Common Good
Edited by Beth Schaefer Caniglia, Beatrice Frank, John L. Knott Jr., Kenneth S. Sagendorf, Eugene A. Wilkerson

Local Activism for Global Climate Justice
The Great Lakes Watershed
Edited by Patricia E. Perkins

Political Responsibility for Climate Change
Ethical Institutions and Fact-Sensitive Theory
Theresa Scavenius

Insuring Against Climate Change
The Emergence of Regional Catastrophe Risk Pools
Nikolas Scherer

Climate Justice and Community Renewal
Resistance and Grassroots Solutions
Edited by Brian Tokar and Tamra Gilbertson

Teaching Climate Change in the United States
Edited by Joseph Henderson and Andrea Drewes

Climate Change Law in China in Global Context
Edited by Xiangbai He, Hao Zhang, and Alexander Zahar

For more information about this series, please visit: www.routledge.com/Routledge-Advances-in-Climate-Change-Research/book-series/RACCR

Climate Change Law in China in Global Context

Edited by
Xiangbai He, Hao Zhang,
and Alexander Zahar

Routledge
Taylor & Francis Group
LONDON AND NEW YORK

earthscan
from Routledge

First published 2020
by Routledge
2 Park Square, Milton Park, Abingdon, Oxon OX14 4RN

and by Routledge
605 Third Avenue, New York, NY 10017

First issued in paperback 2022

Routledge is an imprint of the Taylor & Francis Group, an informa business

British Library Cataloguing-in-Publication Data
A catalogue record for this book is available from the British Library

Library of Congress Cataloging-in-Publication Data
Names: He, Xiangbai, editor. | Zhang, Hao (Law teacher), editor. | Zahar, Alexander, editor.
Title: Climate change law in China in global context / edited by Xiangbai He, Hao Zhang and Alexander Zahar.
Description: Abingdon, Oxon ; New York, NY : Routledge, 2020. | Series: Routledge advances in climate change research | Includes bibliographical references and index.
Identifiers: LCCN 2020006235 (print) | LCCN 2020006236 (ebook) | ISBN 9781138742536 (hardback) | ISBN 9781315182209 (ebook)
Subjects: LCSH: Climatic changes—Law and legislation—China.
Classification: LCC KNQ3130.3 .C57 2020 (print) | LCC KNQ3130.3 (ebook) | DDC 344.5104/6342—dc23
LC record available at https://lccn.loc.gov/2020006235
LC ebook record available at https://lccn.loc.gov/2020006236

ISBN 13: 978-0-367-50970-5 (pbk)
ISBN 13: 978-1-138-74253-6 (hbk)
ISBN 13: 978-1-315-18220-9 (ebk)

DOI: 10.4324/9781315182209

Typeset in Times New Roman
by Apex CoVantage, LLC

Contents

Tables

Contributors

Mandy Meng Fang is a post-doctoral fellow at The Chinese University of Hong Kong. She was previously at the National University of Singapore. She obtained her PhD from CUHK and holds an LLM from University College London. Her research interests include international economic law, climate change governance, and energy law, and her work has been published in journals including the *Journal of World Trade*, the *Journal of World Investment and Trade*, the *Chinese Journal of Environmental Law*, and *Climate Law*.

Xiangbai He is an associate professor at Zhejiang University, Guanghua Law School, China. After receiving her PhD degree from Western Sydney University, she joined Environmental Defense Fund (EDF), Beijing Office, before becoming an academic. Her research interests are primarily in the area of Chinese environmental law and climate change law. In 2016, she published a monograph, *Legal Methods of Mainstreaming Climate Change Adaptation in Chinese Water Management*, with Springer. She has published more than ten articles in both Chinese and English. One of her recent publications on climate change law is "Legal and Policy Pathways of Climate Change Adaptation: Comparative Analysis of the Adaptation Practices in the US, Australia and China", in *Transnational Environmental Law*. She is currently working on an article about China's legal reform of environmental impact assessment.

Ancui Liu is a lecturer in the Faculty of Law at Nankai University, China, where she teaches courses on environmental law. Her doctoral dissertation, from Maastricht University, is entitled *Regulating Genetically Modified Crops in View of Environmental Risks: China's Implementation of International Obligations*. Biosafety law is thus one of her research interests. She is now working on how international experiences with regulating biotechnology can inform China's development of a biosafety law. She is also interested in topics that link biodiversity and climate change. An article she is working on analyses how international instruments are addressing agricultural practices that contribute to the mitigation of greenhouse gas emissions and adaptation to climate change. In addition to Chinese and international environmental law, her research engages European environmental law in comparative analyses.

Ying Shen is a post-doctoral fellow at the Käte Hamburger Kolleg/Centre for Global Cooperation Research at the University of Duisburg-Essen, Germany. She earned her LLB, LLM, and PhD in China (Nanjing University), Germany (University of Göttingen), and Australia (Western Sydney University), respectively. Her expertise lies in the field of environmental law, with a special focus on climate change issues. Dr Shen has been granted several awards and funding grants by the Australian, German, and Chinese governments. She was an Endeavour Research fellow and the first Australia-APEC Women in Research fellow representing China in 2015 and has since held research positions in several universities and institutions in both Australia and Germany. Dr Shen has published articles on climate change issues in environmental law journals in the United States and Australia. She is also the author of *China's Way to Carbon Emissions Reduction: The Choice of Regulatory Instruments and Its Legal Challenges*, a monograph published by Wolters Kluwer in 2015. This was the first legal study to investigate and evaluate policy choices for emission reduction in China in a comprehensive manner.

Alexander Zahar is Honorary Professor at Macquarie Law School, Sydney, Australia, and Professor of Law at Southwest University of Political Science and Law, Chongqing, China. His research interests are in international climate change law as well as public international law and climate change law in China. He is the founder and editor of the journal *Climate Law* (published by Brill/Nijhoff). He periodically serves in Bonn, Germany, as a lead expert reviewer of greenhouse gas inventories, National Communications, and Biennial (Update) Reports submitted by states pursuant to the United Nations Framework Convention on Climate Change. Professor Zahar's most recent monograph, *Climate Change Finance and International Law*, was published by Routledge in 2017 in hardback and issued again in 2018 in paperback. He is currently working on a book edited with Benoit Mayer, *Debating Climate Law* (Cambridge University Press, forthcoming 2021), and on a monograph titled *Climate Law: A Theoretical Introduction.*

Hao Zhang is an assistant professor at the Faculty of Law, The Chinese University of Hong Kong (CUHK Law). He obtained law degrees in China before completing his PhD at Melbourne Law School, University of Melbourne, Australia. His research interests are primarily in the fields of energy law, climate law, and environmental law. He is currently working on the legal issues surrounding the green economic transition in China (i.e. renewable energy and emission trading), and China's energy-sector reform. He has published in leading law and policy journals on these topics. Dr Zhang is the deputy director of the LLM Programme in Energy and Environmental Law at CUHK Law. He is also the associate editor for reviews at the journal *Climate Law*.

Richard Zhang Qing holds a PhD in law from The Chinese University of Hong Kong. His bachelor of laws is from Sun Yat-sen University, Guangzhou, China, and his master's degree (in environmental and natural resources protection law) is from China University of Political Science and Law, in Beijing. He is a member of the Environmental Law Branch of the Chinese Society for

Environmental Sciences and is licensed to practice law in China. His research interests lie in the theory and practice of environmental public-interest lawsuits initiated by NGOs in China. His paper "Public Interest Environmental Litigation under China's Environmental Protection Law" (co-authored with Benoit Mayer) was cited by the United Nations Environment Programme's *Environmental Rule of Law: First Global Report* (January 2019). Dr Zhang Qing's PhD thesis, titled *Ground-Breaking but not so Effective: Environmental Public Interest Litigation Filed by NGOs in China*, received the highest assessment grade. He is an assistant editor at the *Chinese Journal of Environmental Law* and a peer reviewer for *Climate Law*.

Introduction

Xiangbai He, Hao Zhang, and Alexander Zahar

Climate change has been a salient challenge for countries around the globe, and addressing it needs a shared commitment. As the world's largest greenhouse gas (GHG) emitter, China's efforts in combating climate change are crucial for the success of the international climate change regime. This regime is now a quarter-century old, yet little is available in English scholarship about the governance of climate change in the very country that holds the key to the problem's solution.

Most existing scholarship in English has been the work of scholars who do not live in China or are not fluent in Chinese. This can contribute to misunderstandings of China's climate change governance. Furthermore, the work to date has generally been piecemeal and fragmented. This book aims to fill the gap by mapping, analysing, and assessing China's key climate change debates, polices, laws, and regulations. It should help readers who do not know much about the materials in Chinese to develop a better perception of what is really happening in China.

Writing about the climate change laws of any country is challenging. In China, laws and regulations relevant to climate change are not found in a single place but cut across the legal system at multiple levels of government. Some laws enable litigation, so the courts' contribution must also be factored in. Many difficulties of interpretation arise from legal inconsistency in the texts, lax enforcement, and limited public participation.

Seven climate change scholars have contributed to this collection of essays. All of them are Chinese or residents in China. They explain how the country's legal system has been mobilized to support the reduction of GHG emissions, build defences against the impacts of climate change, and meet the country's international obligations. Of course, they not only are interested in China's laws on the books but also seek to explain how they are implemented and integrated, through which institutions, and with what success. China's current laws relating to climate change are certainly an insufficient response to the problem, and many reforms will be necessary in light of the 2015 Paris Agreement.

The book begins with an introduction to the Chinese legal system and institutional framework. This chapter, by Ancui Liu, describes the characteristics, hierarchy of sources of law, and institutions that are related to climate change. It provides a foundation for understanding Chinese laws and policies. The chapter also analyses China's international obligations on climate change and describes corresponding Chinese laws and policies. A range of administrative authorities are involved in dealing with climate change in China, and these institutions are also discussed in this chapter.

Chapter 2, by Xiangbai He, sheds light on the scholarship and debates on climate law and policy in China. An analysis of the scholarly discussion in Chinese sources reveals the key legal concerns related to climate change among Chinese scholars; it also signals China's future regulatory path on climate change. China's climate change strategy is still in transition, from a policy-oriented approach to a legal one; therefore, academic legal writings have relatively few sources of law to base themselves on. The chapter focuses on four specific questions that dominate Chinese legal debates on China's climate law and policy.

The strong growth in GHG emissions in China represents the greatest challenge to the global effort to combat climate change. With the rise of its economy, China has been transformed into a climate superpower, on a par with the European Union and the United States, capable of making or breaking a global climate deal. The role of China in the context of global climate governance is thus of great importance. Chapter 3, by Ying Shen, identifies and discusses the continuities and discontinuities between international climate change law and policy, on the one hand, and China's domestic equivalents, on the other, through a chronological presentation of China's engagement in international climate change negotiations. The first period, from 1989 to 1995, saw China's participation in the early global climate change negotiations leading to the United Nations Framework Convention on Climate Change. The second period, from 1995 to 2007, represented a shift towards more-active participation and engagement with the Kyoto Protocol. The third period started around the time of the Bali Conference in 2007 and marks China's comprehensive engagement with international negotiations in the post-Kyoto era. The chapter, in its conclusion, provides an outlook on China's participation in the global climate change regime.

The development of renewable energy in China is unprecedented. Chapter 4, by Mandy Meng Fang, centres on the effectiveness and challenges of implementing the regulatory mechanisms under China's renewable-energy-law framework. Given China's long-lasting supremacy of a coal-dominated energy structure, the progress of renewable energy in the country has occurred in a relatively short period, mainly due to a proliferation of supportive laws and policies enacted by China's central government over the past two decades. However, it is important to examine the factors impeding the long-term development of renewable energy, attributable to the deficiencies of current renewable-energy laws and policies in China. This chapter points out such problematic issues as the overlapping yet fragmented authority in regulation of renewable-energy development at the central level; incoherence between central and local governments in implementing renewable-energy laws and policies; and a slow revision rate for renewable-energy laws and policies. Solving these problems in order to accelerate the development and deployment of renewable energy to meet increasing energy demands and address climate change is significant for China's future – and for the rest of the world.

As part of the regulatory framework to reduce China's GHG emissions, market-based instruments such as emission trading are part of the mitigation tools that complement the existing command-and-control measures towards decarbonization. For the time being, the market initiatives that underpin China's mitigation strategies include emission trading and a renewable-electricity quota system. Building on existing studies, Chapter 5, by Hao Zhang, examines the legal framework in China for emission trading and other market initiatives that aim to reduce GHG emissions. The framework for emission trading has been gradually developed through a strategy of building a national emission-trading scheme (ETS) through seven local pilot programmes. These pilot ETSs are designed and operated at local levels to test regulatory design and implementation for a national ETS. The renewable-electricity quota system, initiated in 2018, is a relatively new market instrument that mandates priority access of renewable-energy to the grid network through binding targets and trading of renewable energy certificates (RECs). RECs are issued to renewable-energy generators for each megawatt hour they generate, and there are separate obligations for hydropower and other renewable-energy sources (wind, solar, and biomass). The system requires selected participants in China's electricity market, including grid companies, electricity retail companies, and large end-users, to take part in directly trading power by purchasing RECs to prove that a percentage of their electricity comes from renewable energy. The chapter discusses the legal obstacles that hinder the functioning of the two market systems and highlights the interactions between these two instruments as they relate to their effectiveness. This chapter aims to contribute to the scholarly discussion on the interactions between different market-based instruments in China in the context of climate mitigation.

Closely linked to the design of China's national ETS is the inclusion of an offset scheme that provides incentives to increase carbon sinks and sequestration. Chapter 6, also by Hao Zhang, evaluates the institutional and regulatory framework governing the domestic offset scheme in China, in particular the requirements that must be met to qualify the China Certified Emission Reductions (CCERs) for acceptance in China's domestic ETSs. The establishment of China's domestic offset scheme follows on from the governance framework that the Chinese authorities established for participation in the Kyoto Protocol's Clean Development Mechanism (CDM). The central question for this chapter is what drove China to construct its domestic offset scheme and how the scheme's regulatory and institutional framework has been designed. China's offset scheme is particularly important for generating investment in non-ETS sectors that are vulnerable to climate change, such as agriculture and forestry. Still, China's domestic offset regulation is very much a work in progress. Success in moving from participation in the CDM to a national offset scheme is critical to the development of climate law and policy in the country. China will have to navigate several potential pitfalls, especially in ensuring the environmental integrity of offsets.

In its Nationally Determined Contribution (NDC) under the Paris Agreement, China committed to peak its GHG emissions around 2030 and lower the carbon dioxide intensity of GDP by 60%–65% by 2030 from the 2005 level. Under the Paris Agreement and its Rulebook, NDC transparency is critical for accountability among countries, to track global progress, and to understand how commitments and pledges will affect global emissions. Chapter 7, by Alexander Zahar and Hao Zhang, is about China's regulatory system for the monitoring, reporting, and verification (MRV) of GHG emissions. China has reported five national inventories of its GHG emissions. They are for the years 1994, 2005, 2010, 2012, and 2014, and they were published in 2004, 2012, 2018, 2016, and 2018 respectively (but they were not published in chronological order). The increased frequency of publication of China's inventories in recent years is a sign that the country has improved its MRV system for the production of its GHG inventory, pursuant to its international obligations. The objective of this chapter is to describe the MRV system that China has put in place for its inventory of GHGs, including the institutions and data sources that China uses for inventory compilation and quality control of its inventory. The chapter also aims to clarify several MRV-related topics for China, including China's obligations on GHG reporting under the Paris Agreement; the extent of China's use of the law to underpin the design and strengthen the implementation of its climate-related MRV system; and the quality of China's GHG reporting to date. Chapter conclusions include that China's MRV system has recently experienced a significant enhancement and growth in technical expertise by developing its regulatory and institutional capacity and implementing detailed methodological guidelines. However, China's system still needs to incorporate emissions from more sectors of its economy, and its accountability system falls short, with weak penalties and lax enforcement.

Globally, legal frameworks for adaptation to climate change are less developed than those for mitigation, and this also holds true in China. Chapter 8, by Xiangbai He, discusses China's legal and policy framework for adaptation. China has made specialized climate change policies and regulations, in which adaptation is a recognized objective. However, adaptation is inadequately understood in China at present, not least due to the challenges of scientific uncertainty and the location-specific nature of climate change impacts. The current legal and policy framework on adaptation lacks clarity, specificity, and rules of implementation. The chapter argues that, in the interim, adaptation is best integrated into existing development planning and key-project development at the national and subnational levels through the use of the process of environmental impact assessment – although even such an interim measure will not be easy to implement.

Chapter 9, by Richard Zhang Qing, discusses climate change litigation through the lens of environmental public-interest litigation (EPIL) filed by the NGOs in China. The chapter examines whether EPIL cases in China can address climate-change-related concerns directly. It analyses selected EPIL cases to show their potential in this respect. The challenges faced by China's climate change litigation (CCL) – the difficulty of proving the link between emissions and damage and the lack of a court guideline on hearing CCL cases – are explored. The chapter concludes that two NGO-filed cases from 2017 represent a kind of limited breakthrough, by targeting climate change issues directly. However, a dependence on the initiative of plaintiffs in EPIL cases will not be enough to advance China's CCL, given the current lack of supporting mechanisms from both the government and the courts.

1 China's legal system

Sources of law and institutions related to climate change

Ancui Liu

Overview

Since the United Nations Framework Convention on Climate Change (UNFCCC) was concluded in 1992, climate change and its adverse effects have been considered a common concern of humankind.[1] One of the sustainable development goals is to combat climate change and its adverse effects.[2] The Chinese government has shown its ambition in this respect by implementing the UNFCCC and contributing to the finalization of the Paris Agreement.[3] China has also taken measures to deal with climate change domestically.[4] To understand these measures, the first step is to be clear about what China's current laws and policies are and which of them could guide and control activities related to climate change in China.

The literature has analysed current laws and policies related to climate change in China, usually examining the loopholes or legal gaps and giving recommendations on how to improve on them.[5] Many scholars have proposed the adoption of

1 United Nations Framework Convention on Climate Change, United Nations Treaty Series, vol. 1771, p. 107, New York, 9 May 1992, Preamble.

2 'Goal 13: Take Urgent Action to Combat Climate Change and Its Impacts' <www.un.org/sustainabledevelopment/climate-change/> accessed 29 November 2019.

3 For instance, the Clean Development Mechanism (CDM) is one of the financial mechanisms under the Kyoto Protocol. China has participated in CDM projects. See Xi Wang and Tang Tang, 'Research and Scholarship on Climate Change Law in Developing Countries', in Daniel A. Farber and Marjan Peeters (eds), *Climate Change Law,* 121 (Elgar Encyclopedia of Environmental Law, Edward Elgar, 2016). For China's efforts during the negotiation of the Paris Agreement, see Gao Xiaosheng, 'The Paris Agreement and Global Climate Governance: China's Role and Contribution' (2016) 2 *China Quarterly of International Strategic Studies* 378–380; Hongyuan Yu, 'The Paris Climate Agreement and China's Role in Global Climate Governance' (2016) 2 *China Quarterly of International Strategic Studies* 197–198.

4 For domestic efforts that address climate change, see Pu Wang, Lei Liu, and Tong Wu, 'A Review of China's Climate Governance: State, Market and Civil Society' (2018) 18 *Climate Policy* 666–668; Gao (n 3), 376–378.

5 Just to name a few, Yulia Yamineva and Zhe Liu, 'Cleaning the Air, Protecting the Climate: Policy, Legal and Institutional Nexus to Reduce Black Carbon Emissions in China' (2019) 95 *Environmental Science & Policy* 5–7, 9–10; Yen-Chiang Chang, Chuanliang Wang, Mehran Idris Khan, and Nannan Wang, 'Legal Practices and Challenges in Addressing Climate Change

a climate change law as well as suggestions to revise existing legal rules.[6] Local regulations and policies have also been looked at, and it has been shown that local governments are seeking innovative ways to regulate climate change.[7] (For more on the trends in Chinese scholarship in this respect, see the next chapter in this book.)

This chapter is structured as follows. The first section will explain the sources of law in China and their hierarchy. This will provide readers with a background against which to understand China's climate law. The country's international legal obligations on climate change are important parts of Chinese climate change regulations. Therefore, in the second section, the international law that regulates climate change will be discussed. The third section will focus on Chinese laws and policies that address climate change. Lastly, the fourth section will discuss the institutions in China with its responsibilities related to climate change.

Introduction to the Chinese legal system

This section is about the Chinese legal system generally, including its sources of law and their hierarchy. The Chinese legal system and sources of law are different

---

and Its Impact on the Oceans – A Chinese Perspective' (2020) 111 *Marine Policy* 103354; Pan Xiaobin, 'Construction of Climate Change Law System in China' (中国应对气候变化法律体系的构建) (2016) 6 *Nankai Journal (Philosophy, Literature and Social Science Edition)* (南开学报:哲学社会科学版) 78–85; Wang Jian, 'Thoughts on Legal Problems of the Global Response to Climate Change Context' (全球气候变化背景下我国法律应对问题的思考) (2016) 2 *Theory Research* (学理论) 110–112. Alex Wang, 'Climate Change Policy and Law in China', in Cinnamon P. Carlarne, Kevin R. Gray, and Richard Tarasofsky (eds), *Oxford Handbook of International Climate Change Law,* 635–669 (Oxford University Press, 2016). Some publications focus on emission-trading systems, which are important market-based mitigation measures in addressing climate change. See, for example, Coraline Goron and Cyril Cassisa, 'Regulatory Institutions and Market-Based Climate Policy in China' (2017) 17 *Global Environmental Politics* 99–120; Zhe Deng, Dongya Li, Tao Pang, and Maosheng Duan, 'Effectiveness of Pilot Carbon Emissions Trading Systems in China' (2018) 18 *Climate Policy* 992–1011. See also Chapter 2 of the present book.

6 For instance, Yamineva and Liu (n 5), 8–10; Tian Danyu, 'Study of Legislation on Climate Change' (应对气候变化立法研究) (2018) 3 *World Environment* (世界环境) 62; Li Yanfang, Zhang Zhongli, and Li Cheng, 'Toward Specialized Legislation to Address Climate Change in China' (我国应对气候变化立法的若干思考)' (2016) 33 *Journal of Shanghai University (Social Sciences)* (上海大学学报: 社会科学版) 8–11. Comparative studies have also been carried out. See Xiangbai He, 'Legal and Policy Pathways of Climate Change Adaptation: Comparative Analysis of the Adaptation Practices in the United States, Australia and China' (2018) 7 *Transnational Environmental Law* 347–373; Isabella Neuweg and Alina Averchenkova, 'Climate Change Legislation and Policy in China, the European Union and the United States', in A. Averchenkova, S. Fankhauser, and M. Nachmany (eds), *Trends in Climate Change Legislation,* 37–59 (Edward Elgar, 2017).

7 Miranda Schreurs, 'Multi-Level Climate Governance in China' (2017) 27 *Environmental Policy and Governance* 169–171; Pan Xiaobin, 'Research on Climate Change Prior Legislation of Chinese Local Government' (中国地方应对气候变化先行立法研究) (2017) 38 *Law Science Magazine* (法学杂志) 136–137. This mode is summarized by some scholars as a bottom-up approach. See Schreurs (n 7), 169; Pan (n 7), 133; Peter H. Koehn, *China Confronts Climate Change: A Bottom-up Perspective,* 1–2 (Routledge, 2015).

from those of other countries. Terms such as *laws* and *regulations* do not have the same meanings in the Chinese legal context as they do elsewhere. From a typological perspective, Chinese laws and policies related to climate change can be better understood once they are categorized into different sources of law.

Character of the Chinese legal system

The Chinese legal system is not a pure form of a civil law system, although it shares features with that system. It is often described in the literature as a socialist legal system with Chinese characteristics.[8] An important characteristic of the Chinese legal system is the leadership of the Communist Party of China.[9] The enactment and implementation of laws are heavily influenced by Party policy. For some new emerging issues, the use of policies is regarded as an appropriate tool.[10] For instance, policies on emission-trading systems and low-carbon cities came into existence before laws on those topics did.[11] A policy may be transformed into legislation after it has been tried out for a while.[12] The emphasis of the Party has been shifting from economic development to environmental protection.[13] The balance of these two interests should therefore be considered when Chinese laws and policies as a response to climate change are analysed.

In 2008, a government White Paper on the rule of law divided the Chinese legal system into seven branches: constitutional law, civil and commercial law, administrative law, criminal law, economic law, social law, and litigation and non-litigation procedural law.[14] The country's Constitution and a statute known as the Legislation Law provide rules and procedures for the enactment and interpretation

8 Roderick O'Brien, 'Comparative Law and China's Socialist Legal System' (2014) 2014 *Journal of South African Law* 130; Xin Chunpeng, 'The Socialist Legal System with Chinese Characteristics and Its Significance' (中国特色社会主义法律体系及其重大意义) (2014) 36 *Chinese Journal of Law* (法学研究) 19–26; Mo Zhang, 'The Socialist Legal System with Chinese Characteristics: China's Discourse for the Rule of Law and a Bitter Experience' (2010) 24 *Temple International & Comparative Law Journal* 2–3, 48–49.

9 Li Lin, *Building the Rule of Law in China,* 218 (Elsevier Asian Studies Series, Chandos Publishing, 2017); Zhang (n 8), 50.

10 Fang Dongsheng, 'Recomprehension of Relation between Law and Policy – Empiricism as the Angle of View' (法律与政策关系的再认识 – 以实证为视角) (2008) 29 *Journal of Jining University* (济宁学院学报) 87–89; Fang Shirong and Sun Caihua, 'Reconsideration of the Relationship between Policies and Law – from the Viewpoint of Concentrating and Reflecting the Will of the People' (关于政策与法律关系的再思考 – 从集中和反映人民意志的视角) (2007) 5 *Hubei Social Sciences* (湖北社会科学) 17, 19.

11 Pan (n 5), 80–81; Li, Zhang, and Li (n 6), 3–4.

12 Lin (n 9), 158–159.

13 This is reflected in the Party's views on climate change. Before 2007, climate change was perceived basically as an economic issue that would slow down economic development. Since 2007, climate change has been regarded as a sustainable development issue. See Maria Julia Trombetta, 'Securitization of Climate Change in China: Implications for Global Climate Governance' (2019) 5 *China Quarterly of International Strategic Studies* 106–109.

14 State Council Information Office, China, 'China's Efforts and Achievements in Promoting the Rule of Law' (2008) 7 *Chinese Journal of International Law* 520–521.

of laws in China.[15] The development of the Chinese legal system has progressed steadily.[16] However, new developments, such as climate change, cause it to lag behind in some respects. The Standing Committee of the National People's Congress (NPC) considers the drafting of climate-change-related laws and regulations to be an important step in improving the Chinese legal system.[17] In 2009, it passed a resolution to enact a climate change law to strengthen the national response to climate change.[18] China is planning to enact laws (*fǎlǜ*) and regulations that address climate change, subject to China's National Plan on Climate Change.[19] However, at the time of writing, no specific climate change law had been enacted.[20]

Sources of law in China

Sources of law can be either formal or informal.[21] A definition of formal sources is that they are "available in an articulated textual formulation embodied in an authoritative legal document".[22] Informal sources of law can be defined as "legally

15 Jianfu Chen, *Chinese Law: Context and Transformation (Revised and Expanded Edition)*, 239 (Martinus Nijhoff Publishers, 2016); Yan Lin, 'Constitutional Evolution through Legislation: The Quiet Transformation of China's Constitution' (2015) 13 *International Journal of Constitutional Law* 72–73.

16 Compared with the legal system in 1970s, which hardly provided any law, there are now laws to go by or regulations to follow. See 'Socialist System of Laws Established in China' <www.china.org.cn/china/NPC_CPPCC_2011/2011-03/10/content_22099470.htm> accessed 29 November 2019. Albert H. Y. Chen, 'China's Long March Towards Rule of Law or China's Turn against Law?' (2016) 4 *Chinese Journal of Comparative Law* 11.

17 Resolution of the Standing Committee of the National People's Congress on Making Active Responses to Climate Change (全国人民代表大会常务委员会关于积极应对气候变化的决议) Standing Committee of the National People's Congress, 27 August 2009, Section 4. Some scholars have analysed the necessity to enact a climate change law in China. See, for instance, Tian (n 6), 59–60; Li Yanfang, 'On the Establishment of the Legal System of China's Response to Climate Change' (论中国应对气候变化法律体系的建立) (2010) *Journal of China University of Political Science and Law* (中国政法大学学报) 84–85.

18 Resolution of the Standing Committee of the National People's Congress on Making Active Responses to Climate Change (n 17).

19 Chapter 7, section 1, NDRC, National Plan on Climate Change (2014–2020) (国家应对气候变化规划(2014–2020年)) <www.ndrc.gov.cn/zcfb/zcfbtz/201411/W020141104584717807138.pdf> accessed 29 November 2019.

20 A draft proposal was released for public comment in March 2012. The draft proposal was developed by such institutions as the Institute of Law of the CASS and the Swiss Agency for Development and Cooperation and then submitted to the NDRC. Although the NDRC attempted to coordinate all relevant parties to reach a consensus regarding the proposed climate change law, no report was released in relation to the submission of a draft climate change law from the NDRC to the State Council or the NPC. For the draft proposal, see Chinese Academy of Social Sciences Institute of Law and the Swiss Agency for Development and Cooperation, 'The Act of the People's Republic of China Addressing Climate Change (the Draft Proposal)' (2014) 2 *China Legal Science* 122–159. Zhou Chen explains the reasons for, and main aspects of, the proposed climate change law. See Chen Zhou, *The Legal Barriers to Technology Transfer under the UN Framework Convention on Climate Change*, 169–170 (Springer, 2019).

21 Edgar Bodenheimer, *Jurisprudence: The Philosophy and Method of the Law*, 325 (Harvard University Press, 1981).

22 Ibid. Examples of the formal sources of law include constitutions, statutes, executive orders, administrative regulations, and judicial precedents.

significant materials and considerations which have not received an authoritative or at least articulated formulation and embodiment in a formalized legal document".[23] Thus, according to Bodenheimer, informal sources are subsidiary to formal ones, such that when the formal sources are explicitly stated, the informal ones, in most cases, need not or should not be consulted.[24] However, when the formal sources are ambiguous, the informal ones may be resorted to or relied on.[25]

Here, I follow the definitions of the formal and informal sources of law provided by Bodenheimer. The formal sources of law in China include the Constitution, laws (*fǎlǜ*), regulations, administrative measures, international law, and judicial interpretations.[26] The informal sources include policies, judgments, and so-called other normative documents. Each source of law will be briefly explained below. Policies sometimes have greater practical influence than formal sources of law as a result of the special role and mode of operation of the government in China.[27] A discussion of policies is therefore included in this section.

Constitution

China's Constitution was amended in 2018, after four earlier amendments (in 1988, 1993, 1999, and 2004).[28] The Constitution contains a preamble and four chapters, consisting of 143 articles.[29] The Constitution contains provisions in relation to the government's powers and citizens' rights and interests. Substantive rules in the Constitution function as declaratory rules, because judges cannot interpret constitutional rules.[30] The Constitution and Law Committee, a specialized

23 Ibid. Examples of informal sources of law, according to Bodenheimer, include standards of justice, individual equity, public policies, and customary law.

24 Ibid.

25 Ibid., 325–326.

26 Terms, such as *laws* (*fǎlǜ*) and *regulations* (*tiáolì*), are used to define the scope of application and the level of authority of enactment. See James M. Zimmerman, *China Law Deskbook: A Legal Guide for Foreign-Invested Enterprises*, 65–66 (3rd edn, Section of International Law, American Bar Association Book Publishing, 2010).

27 Huizhen Chen, *Towards a Market-Based Climate Regime in China?: A Legal Perspective on the Design and Implementation of Greenhouse Gas Emissions Trading* (Doctoral Dissertation, Maastricht University, 2015) 135.

28 The PRC's first constitution was adopted in 1954. The current constitution was adopted in 1982. For an introduction to the Constitution, see Paul Kossof, *Chinese Legal Research,* 14–15 (International Legal Research Series, Carolina Academic Press, 2014).

29 Chapter one of the Constitution sets out general principles. Chapter two provides for the fundamental rights and duties of citizens. Chapter three lays down rules on state organs. And chapter four is about the national flag, anthem, emblem, and the capital.

30 Lin (n 15), 89; Chuan Feng, Leyton P. Nelson, and Thomas W. Simon, *China's Changing Legal System: Lawyers & Judges on Civil & Criminal Law,* 13 (Palgrave Macmillan, 2016); Qianfan Zhang, 'A Constitution without Constitutionalism? The Paths of Constitutional Development in China' (2010) 8 *International Journal of Constitutional Law* 952. Although there is an absence of judicial review, the Constitution produces some beneficial effects. One reason for the government to respect the constitutional rules and citizens' rights and interests is that the government wants to maintain a good image. See Zhang (n 30), 957–960.

committee of the NPC, is responsible for constitutional review and for the implementation and interpretation of the Constitution.[31] By carrying out constitutional reviews, the committee examines whether laws are consistent with the Constitution. However, the rules and procedures for constitutional review have not yet been elaborated.[32]

Laws

The term *laws* (*fǎlǜ* in Chinese) has different meanings in the Chinese legal system.[33] In a narrow sense, it refers to the body of legislation enacted by the NPC and its Standing Committee, as prescribed by Articles 62(3), 67(2), and 67(3) of the Constitution.[34] *Laws* also has broad sense, encompassing all types of legislation, regulation, and administrative measures.[35] In this chapter, to avoid confusion, *laws* in the narrow sense will be followed by *fǎlǜ* in parentheses.

Regulations

There are two types of regulations. One is administrative regulations and the other is local regulations. Administrative regulations and local regulations, in accordance with the Constitution and laws (*fǎlǜ*), are enacted, respectively, by the State Council and certain levels of local people's congresses (more below). Local regulations cannot conflict with administrative regulations.[36]

The State Council is the executive arm of China's government and the highest administrative authority in China. It is responsible for the enactment of administrative regulations,[37] which it passes subject to the Constitution and the delegation of legislative authority from the NPC and its Standing Committee.[38]

31 The Constitution of the People's Republic of China (中华人民共和国宪法) National People's Congress, adopted on 4 December 1982, amended on 12 April 1988, 29 March 1993, 15 March 1999, 14 March 2004, and 11 March 2018, art. 70. Standing Committee of the National People's Congress, Decision of the Standing Committee of the National People's Congress on Matters concerning the Duties of the Constitution and Law Committee of the National People's Congress, 22 June 2018, para. 2.

32 Some scholars have made suggestions about how to carry out constitutional review. See Yu Wenhao, 'Discussions on Duties of the Constitution and Law Committee on Constitutional Review' (宪法和法律委员会合宪性审查职责的展开) (2018) *China Legal Science* (中国法学) 61–63; Zou Pingxue, 'Missions and Institutional Innovation of the Constitution and Law Committee' (宪法和法律委员会的目标定位与机制创新) (2018) 4 *China Law Review* (中国法律评论) 47–50.

33 Lin (n 9), 220.

34 Zimmerman (n 26), 65–66.

35 Briefly, regulations can be enacted by the State Council and local people's congresses, and administrative measures can be enacted by ministries, commissions, and local governments.

36 The Legislation Law of the People's Republic of China (中华人民共和国立法法) National People's Congress, adopted on 15 March 2000, amended on 15 March 2015, art. 88, para. 2.

37 Constitution (n 31), art. 89, para. 1.

38 Chen (n 15), 242–243.

Local people's congresses and local people's governments are divided into four levels: province, prefecture, county, and township.[39] This division follows the administrative division of regions.[40] Local people's congresses and their standing committees – for provinces, autonomous regions, municipalities directly under the central government, cities divided into districts, cities where special economic zones are located, and cities deemed by the State Council as major cities – have the authority to make local regulations.[41] These can be adopted when laws (*fǎlǜ*) provide for it or when local governments deem it necessary to do so on the basis of regional circumstances.[42] There might be different local regulations addressing the same subject matter, as each level of local government mentioned earlier is given equal power to enact local regulations. However, not all local people's congresses have such authority. For example, local people's congresses for counties are not given authority to adopt local regulations.

Administrative measures

Administrative measures may be categorized as ministerial administrative measures and local administrative measures. Ministerial administrative measures are enacted by ministries and commissions within the central government, while local administrative measures are enacted by certain levels of local government.[43]

Ministries and commissions are under the direction of the State Council.[44] Each is responsible for a particular policy area. The head of each ministry or commission is a member of the State Council.[45] Within each ministry or commission are several divisions and offices. People's governments of provinces, autonomous regions, municipalities directly under the central government, capital cities

39 Constitution (n 31), art. 2, para. 2, and art. 95, para. 1.

40 Administrative Division <http://english.gov.cn/archive/china_abc/2014/08/27/content_281474983873401.htm> accessed 29 November 2019.

41 Constitution (n 31), art. 100; Legislation Law (n 36), art. 72. In addition to local regulations, people's congresses of autonomous regions can make autonomous regulations (*zìzhì tiáolì*) and separate regulations (*dānxíng tiáolì*), depending on their political, economic, and cultural circumstances. See Legislation Law (n 36), art. 75; Chen (n 15), 246–248.

42 Daniel C. K. Chow, *The Legal System of the People's Republic of China in a Nutshell*, 158–159 (3rd edn, West Academic Publishing, 2015); Kossof (n 28), 30–31. Provincial people's congresses and their standing committees can make local regulations, provided that they do not violate the Constitution, laws (*fǎlǜ*), and administrative regulations. See Lin (n 15), 72.

43 Constitution (n 31), art. 90, para. 2, and art. 107; Legislation Law (n 36), arts. 80 and 82.

44 Constitution (n 31), art. 89, para. 3. Currently, there are a total of 27 ministries and commissions in the State Council, which are composed of 21 ministries, three commissions, the National Audit Office, the People's Bank of China, and the State Council's General Office. See State Council Institutional Reform Plan of 2018 (2018年国务院机构改革方案) National People's Congress, effective as of 17 March 2018.

45 The State Council is composed of a premier, several vice-premiers, state councillors, ministers in charge of ministries and commissions, the auditor-general, and the secretary-general. See Constitution (n 31), art. 86, para. 1; State Council Organization Chart <http://english.gov.cn/state_council/2014/09/03/content_281474985533579.htm> accessed 29 November 2019.

of provinces and autonomous regions, cities where special economic zones are located, and cities deemed by the State Council as major cities may enact local administrative measures.[46] Other parts of local governments, such as people's governments of counties, are not allowed to adopt local administrative measures. There might be different local administrative measures regarding the same subject matter, as each level of local government mentioned earlier is given equal power to enact local administrative measures.

International law

International law is recognized as a source of law in the Chinese legal system.[47] Some legislation provides that international law prevails over domestic rules that are inconsistent with international law.[48] For example, Article 142(2) of the General Principles of the Civil Law requires that where an international treaty concluded or acceded to by China contains provisions differing from those in the civil laws of China, the provisions of the international treaty are to apply unless the provisions are ones on which China has announced reservations.[49]

China takes a dualistic approach to the domestic implementation of international law.[50] By taking this approach, international instruments are not implemented automatically in China upon ratification.[51] Furthermore, there are two types of international rules for China: self-executing and non-self-executing.[52] For the former, no new domestic legislation is needed.[53] For example, according to Article 142(2) of the General Principles of the Civil Law, international treaties that contain civil law rules will be applied in China without special measures in Chinese law. For non-self-executing rules, China must incorporate the

46 Legislation Law (n 36), art. 82.

47 John Bahrij and Lily Ko, 'An Overview of English Resources for Chinese Legal Research' (2013) 13 *Legal Information Management* 26. And Chow (n 42), 172–173:

> Although there is no mention of international law as a source of law in the PRC Constitution, scholars in the PRC point out that under the prevailing theory of international law and recent legislative practice in the PRC, international law is part of the domestic law of the PRC and where there is a conflict between a rule of international law contained in a treaty or in customary international law and PRC local law, the rule of international law should prevail.

48 Xi Wang, *Environmental Law in China,* 52 (Kluwer Law International, 2012); Chow (n 42), 173.

49 General Principles of the Civil Law of the People's Republic of China (中华人民共和国民法通则) National People's Congress, adopted on 12 April 1986, amended on 27 August 2009, art. 142, para. 2. Article 260 of the Law of Civil Procedure has a similar provision.

50 Feng, Nelson, and Simon (n 30), 186.

51 Lal Kurukulasuriya and Nicholas A. Robinson, *Training Manual on International Environmental Law*, 20 (UNEP/Earthprint, 2006).

52 Chow (n 42), 173.

53 Kurukulasuriya and Robinson (n 51), 20.

international law into its domestic law once the NPC or its standing committee has ratified that law.[54]

Judicial interpretations and judgments

Sources of law have different roles in court trials.[55] Laws (*fǎlǜ*) form the basis of a court judgment. Other types of legislation, including regulations and administrative measures, may be referred to in judgments. Judicial interpretations are adopted by the Supreme People's Court (SPC) to address questions that arise by the lower courts in the application of laws or to supplement or elaborate existing laws (*fǎlǜ*).[56] Judicial interpretations have the same legal force as laws (*fǎlǜ*), in the sense that both are important bases for decision-making.[57] The SPC also issues opinions and guidance documents to guide lower courts in their trials.[58]

Judgments are not a formal source of law in China.[59] In other words, they have no legal force over subsequent judgments and do not establish a precedent.[60] To make court judgments more consistent with each other, China established a case-guidance system.[61] The SPC selects and declares guiding cases.[62] These cases have a de facto binding effect on lower-level courts: although they are not a

54 Chow (n 42), 173; Guo Dongmei, 'Analysis of Legal Strategy against Climate Change: From Outer System to Inner Rules' (气候变化法律应对实证分析 – 从国际公约到国内法的转化) (2010) 12 *Journal of Southwest University of Political Science & Law* (西南政法大学学报) 45–46.

55 Liu Songshan, 'The Criterion the Court Relies on for Trial' (人民法院的审判依据) (2006) 24 *Tribune of Political Science and Law: Journal of China University of Political Science and Law* (政法论坛: 中国政法大学学报) 22.

56 Legislation Law (n 36), art. 104; Chow (n 42), 179–180; Feng, Nelson, and Simon (n 30), 59–60.

57 Miao Yan, 'Jurisprudential Reflection and Structure Optimization of the Judicial Interpretation System' (司法解释制度之法理反思与结构优化) (2019) 25 *Law and Social Development* (法制与社会发展) 96–97.

58 Chow (n 42), 180–181; Feng, Nelson, and Simon (n 30), 60–62.

59 Bahrij and Ko (n 47), 26. The court does not hold a central position in the Chinese legal system as it does in the common-law system. See Chow (n 42), 174–175.

60 Kossof (n 28), 29; Bahrij and Ko (n 47), 26; Zimmerman (n 26), 69–70.

61 The case-guidance system, which was established under the Provisions of the Supreme People's Court Concerning Work on Case Guidance, refers to the process of selecting, releasing, and using guiding cases throughout the country. See Provisions of the Supreme People's Court Concerning Work on Case Guidance (关于案例指导工作的规定) Supreme People's Court Order No. (2010)51, 26 November 2010, art. 1.

62 For the procedure of collecting, selecting, and releasing guiding cases, see Provisions of the Supreme People's Court Concerning Work on Case Guidance (n 61), art. 4 to art. 13. See also Mark Jia, 'Chinese Common Law? Guiding Cases and Judicial Reform' (2016) 129 *Harvard Law Review* 2221–2224; Bahrij and Ko (n 47), 26; Zimmerman (n 26), 70. Guiding cases are published on the website of the Supreme People's Court. One hundred twelve guiding cases had been published at the time of writing. See Guiding Cases, Official Website of the Supreme People's Court of the People's Republic of China, <www.court.gov.cn/shenpan-gengduo-77.html> accessed 29 November 2019.

binding precedent, they serve as an important reference for lower-level courts in their judgments.[63]

"Other normative documents"

Normative documents include laws (*fălǜ*), regulations, and administrative measures.[64] "Other normative documents" refers to documents that have general binding force that affects the rights and duties of citizens, legal persons, or organizations.[65] "Other normative documents" are issued by administrative authorities, including the State Council and its ministries and commissions and local governments and their departments.[66] Roughly speaking, "other normative documents" are subordinate to formal sources of law, even though they are adopted by administrative authorities and have general binding force.[67]

Hierarchy of sources of law in China

The Constitution and the Legislation Law provide for the general rules on the hierarchy of sources of law in China.[68] The Constitution holds the highest place in the hierarchy and is the core of the Chinese legal system. Laws (*fălǜ*) are the next highest in the hierarchy.[69] Regulations and administrative measures follow in third and fourth place, respectively. Sources of law at a higher level may revise or repeal those at a lower level, provided certain circumstances are met.[70]

There are at least five scenarios: First, the NPC may alter or annul enacted laws (*fălǜ*) and decisions made by its standing committee, as it deems appropriate.[71] Second, the Standing Committee of the NPC may annul administrative regulations, decisions, and orders adopted by the State Council if they contravene the Constitution or laws (*fălǜ*).[72] The Standing Committee also has the

63 Provisions of the Supreme People's Court Concerning Work on Case Guidance (n 61), art. 7; Jia (n 62), 2232–2233. Guiding cases are not sources of law in a formal sense. See Jia (n 62), 2232; Dahrij and Ko (n 47), 26.

64 Wang Xiahao, 'Study on Statute as a Branch of Legal Resource System' (论作为法的渊源的制定法) (2017) 35 *Tribune of Political Science and Law* (政法论坛) 51.

65 Chen (n 15), 255.

66 Ibid. Wen Hui, 'Study on the Recording and Review System of Governmental Normative Documents' (政府规范性文件备案审查制度研究) (2015) 36 *Law Science Magazine* (法学杂志) 10.

67 Wen (n 66), 10; Huang Jinrong, 'Legal Definition and Legal Nature of Normative Documents' ("规范性文件" 的法律界定及其效力) (2014) 7 *Legal Science* (法学) 12–13; Wang Kai, 'Discussion on the Recording of Normative Documents' (论规范性文件的备案审查) (2010) 2010 *Zhejiang Social Sciences* (浙江社会科学) 11.

68 Chen (n 15), 249–254.

69 Ibid., 241–242.

70 Ibid., 248; Chow (n 42), 161.

71 Constitution (n 31), art. 62, para. 12; Legislation Law (n 36), art. 97, para. 1.

72 Constitution (n 31), art. 67, para. 7; Legislation Law (n 36), art. 97, para. 2.

power to annul local regulations adopted by local people's congresses at the provincial level if they contravene the Constitution, laws (*fǎlǜ*), or administrative regulations.[73] Third, local people's congresses have the power to revise or annul legislation or "other normative documents" adopted by their standing committees.[74] For example, a people's congress at the provincial level may annul local regulations that have been enacted by its standing committee.[75] Fourth, standing committees of local people's congresses have the power to annul legislation or "other normative documents" enacted by lower-level people's congresses or local people's governments at the same level.[76] For example, a standing committee of a provincial people's congress can annul local administrative measures enacted by its provincial people's government. Fifth, the State Council has the power to revise or annul administrative measures or "other normative documents" made by ministries or commissions directly under the State Council and local governments.[77]

Furthermore, the Legislation Law provides for ways to solve conflicts between regulations and administrative measures. First, ministerial administrative measures are supplementary to administrative regulations.[78] Second, ministerial administrative measures and local administrative measures might regulate the same topic with different legal provisions. Ministerial administrative measures and local administrative measures have equal legal force.[79] The State Council is to decide on which administrative measures are to be used when there is a conflict.[80] Third, the State Council is to decide which rules to apply when there is a conflict between ministerial administrative measures and local regulations.[81] If the State Council decides that local regulations should be used, the local regulations are to be used directly.[82] But if the State Council decides that ministerial administrative measures should be used, the dispute is to be submitted to the Standing Committee of the NPC to make the final decision.[83]

73 Constitution (n 31), art. 67, para. 8; Legislation Law (n 36), art. 97, para. 2.
74 Constitution (n 31), art. 99, para. 2. Chow (n 42), 161.
75 Legislation Law (n 36), art. 97, para. 4.
76 Constitution (n 31), art. 104; Legislation Law (n 36), art. 97, para. 5.
77 Constitution (n 31), art. 89, para. 13 and art. 89, para. 14; Legislation Law (n 36), art. 97, para. 3. Chow (n 42), 161. Similarly, people's governments at the provincial level shall alter or annul administrative measures made by the people's governments at a lower level that they deem inappropriate. See Legislation Law (n 36), art. 97, para. 6.
78 Chen (n 15), 242.
79 Legislation Law (n 36), art. 91.
80 Legislation Law (n 36), art. 95, para. 3. Nonetheless, the Legislation Law does not provide practical guidelines as to how to apply ministerial administrative measures and local administrative measures when they both are required to address one issue. See Chen (n 15), 255.
81 The Legislation Law lays down the relationship between ministerial administrative measures and local regulations. See Legislation Law (n 36), art. 95; Chow (n 42), 162. However, the practicability of this rule is questionable. See Chen (n 15), 253–254.
82 Legislation Law (n 36), art. 95.
83 Ibid.

Local rules are valid provided that they do not breach laws (*fǎlǜ*) or administrative regulations and are enacted under the authority of local people's congresses or governments.[84]

Importance of Party and government policies

The definition of *policy* is not clear in the Chinese legislation or legal literature.[85] China has Party policies and government policies.[86] Policies differ from laws in several aspects, such as in their initiators, procedures of adoption, scope of application, legal effect, and continuity.[87] Laws have advantages over policies from the perspective of legal certainty, the nature of normative rules, and general application. Moreover, laws outweigh policies, in the sense that policies are informal sources of law, while legislation is a formal source. Nonetheless, policies can provide experience leading up to the enactment and implementation of laws.[88] For instance, laws on emission-trading schemes have been preceded by policies.[89]

A Plan for National Economic and Social Development, which is also referred to as a Five-Year Plan, is an example of a policy instrument. A Five-Year Plan is prepared every five years by the State Council and approved by the NPC on the basis of Party recommendations.[90] A Five-Year Plan sets out targets and guidelines on social and economic development. The plan current at the time of writing – the 13th Five-Year Plan – was adopted in March 2016.[91] To implement a Five-Year Plan, the State Council and its ministries and commissions make more-specific plans (i.e. policies). Although such policies are not formal sources of law, they are

84 The Constitution (n 31), arts. 99 and 100. The Legislation Law (n 36), art. 96. "As long as *fagui* issued below provincial level have the approval of higher authorities, they are valid local rules." From Chen (n 15), 246.

85 Zhang Hong, 'State Policies as a Source of Civil Law' (2017) 38 *Social Sciences in China* 68.

86 Zheng Jinggao and Tian Ye, 'From "State Will" to "Rule of Law" – Extensive Viewpoints of Law on the Relationship between Law and Policy' (从"国家意志"到"行政法治" – 在法律与政策关系上的泛法律观) (2007) 7 *Journal of China University of Geosciences (Social Sciences Edition)* (中国地质大学学报:社会科学版) 100; Liu Zuoxiang, 'Norm Systems in Contemporary China: Theory and Institutional Structure' (当代中国的规范体系:理论与制度结构) (2019) 7 *Social Sciences in China* (中国社会科学) 96–97.

87 Xiao Jinming, 'Restructuring the Relationship of Law and Policy for the Purpose of a Comprehensive Rule of Law' (为全面法治重构政策与法律关系) (2013) 5 *Chinese Public Administration* (中国行政管理) 38–39. Zheng and Tian (n 86), 100.

88 Liu (n 86), 94, 99–100. Xing Huiqiang, 'Growth of Policies and Emptiness of Law: From the Perspective of Economic Law' (政策增长与法律空洞化--以经济法为例的观察) (2012) 3 *Law and Social Development* (法制与社会发展) 120–121. Chen Tan, 'A Primary Discussion on Legitimation and Enactment of Policy' (浅论政策合法化与政策法律化行政与法) (2001) 1 *Public Administration & Law* (行政与法) 55.

89 Marjan Peeters, Huizhen Chen and Zhiping Li, 'Contrasting Emission Trading in the EU and China: An Exploration of the Role of the Courts' (2016) 6 *Climate Law* 221.

90 The Constitution (n 31), art. 62, para. 10 and art. 89, para. 5.

91 The Thirteenth Five-Year Plan for National Economic and Social Development of the People's Republic of China (国民经济和社会发展第十三个五年规划纲要) National People's Congress, adopted on 16 March 2016.

binding, in the sense that they provide targets and guidelines for national authorities and cadres.[92]

China's obligations under international law related to climate change

China's international obligations on climate change are found in the UNFCCC (1992), Kyoto Protocol (1997), and Paris Agreement (2015). Enacting and implementing laws and policies in China and establishing institutions that deal with climate change are normally related to efforts to combat climate change under the international law related to climate change.[93] This section will discuss the history and the main mechanisms of the three international instruments and analyse China's actions thereunder.

UNFCCC

The UNFCCC was adopted in May 1992. It entered into force on 21 March 1994. As of November 2019, it has 197 parties. China signed it on 11 June 1992 and ratified it on 5 January 1993. The parties to the UNFCCC aim to stabilize GHG concentrations in the atmosphere "at a level that would prevent dangerous anthropogenic interference with the climate system".[94] GHGs are defined as "those gaseous constituents of the atmosphere, both natural and anthropogenic, that absorb and re-emit infrared radiation".[95] Legal principles prescribed by the UNFCCC include equity and common but differentiated responsibilities and respective capabilities,[96] the precautionary principle,[97] sustainable development,[98] and international cooperation.[99]

UNFCCC obligations are of two kinds: one is for all parties, and the other is for developed-country parties (so-called Annex I parties).[100] All parties are to formulate, implement, publish, and regularly update national programmes containing measures to mitigate climate change and measures to facilitate adequate adaptation to climate change;[101] moreover, all parties are to take account of climate change

92 Xiao (n 87), 38. For the relationship between the CPC and the Chinese administrative authorities, see Chow (n 42), 145.

93 Chapter 3 of this book will discuss how international law related to climate change influences, or is influenced by, Chinese laws and policies related to climate change.

94 UNFCCC (n 1), art. 2.

95 Ibid., art. 1, para. 5.

96 Ibid., art. 3, para. 1, and art. 3, para. 2.

97 Ibid., art. 3, para. 3.

98 Ibid., art. 3, para. 4.

99 Ibid., art. 3, para. 5.

100 Lavanya Rajamani, 'The United Nations Framework Convention on Climate Change: A Framework Approach to Climate Change', in Daniel A. Farber and Marjan Peeters (eds), *Climate Change Law*, 207 (Elgar Encyclopedia of Environmental Law, Edward Elgar, 2016).

101 UNFCCC (n 1), art. 4, para. 1(b).

considerations, to the extent feasible, in their social, economic, and environmental policies and actions.[102] On the other hand, Annex I parties are required to adopt national policies and take corresponding measures on the mitigation of climate change.[103] This approach embodies the principle of common but differentiated responsibilities. Many obligations are general and abstract, leaving great discretion to the parties.

Article 4(1) of the UNFCCC lists the obligations of developing countries, which include China:

(a) Develop, periodically update, publish, and make available to the Conference of the Parties (COP) national inventories of anthropogenic emissions by sources and removals by sinks of GHGs.
(b) Formulate, implement, publish, and regularly update national and, where appropriate, regional programmes containing measures to mitigate climate change and measures to facilitate adequate adaptation to climate change.
(c) Promote and cooperate in the development, application, and diffusion of technologies, practices, and processes that control, reduce, or prevent GHG emissions in all relevant sectors.
(d) Promote sustainable management, and promote and cooperate in the conservation and enhancement, as appropriate, of sinks and reservoirs of GHGs.
(e) Cooperate in preparing for adaptation to the impacts of climate change.
(f) Take climate change considerations into account, to the extent feasible, in relevant social, economic, and environmental policies and actions, and employ appropriate methods, for example impact assessments with a view to minimizing adverse effects on the economy.
(g) Promote and cooperate in scientific, technological, technical, socio-economic, and other research, exchange relevant scientific, technological, technical, socio-economic, and legal information, enhance education, training, and public awareness related to climate change, and encourage the widest participation in this process.
(h) Communicate to the COP information related to implementation of this convention

As mentioned earlier, China has not enacted a stand-alone climate change law. This to some extent affects its efficient implementation of the UNFCCC.[104] Since 2007, the country has had a National Climate Change Programme, which was

102 Ibid., art. 4, para. 1(f). Peter-Tobias Stoll, 'The Climate as a Global Commons', in Daniel A. Farber and Marjan Peeters (eds), *Climate Change Law,* 136 (Elgar Encyclopedia of Environmental Law, Edward Elgar, 2016).

103 UNFCCC (n 1), art. 4, para. 2(a).

104 This does not mean that enacting a stand-alone climate change law is the only way to implement the UNFCCC. However, gaps in the Chinese legal system in relation to climate change will be overcome by enacting a climate change law. See Guo (n 54), 50.

prepared by the National Development and Reform Commission (NDRC) and adopted by the State Council.[105] It is one of the first national climate change programmes prepared by a developing country.[106] It sets several climate change targets to be achieved by 2010. One such target is that China will lower its carbon dioxide (CO_2) emissions per unit of GDP by 20% from the 2005 level by 2010 and increase forest coverage by 20% from the same level.[107] Other targets include increasing the capacity for adaptation, encouraging scientific research and development of technologies, promoting public awareness, and enhancing administrative capacity.[108]

In 2009, the Chinese government announced that, by 2020, it would lower CO_2 emissions per unit of GDP by 40%–45% from the 2005 level; increase the share of non-fossil fuels in primary energy consumption to about 15%; and increase the forested area by 40 million hectares and the forest-stock volume by 1.3 billion cubic metres compared to the 2005 level.[109] In response to its obligations under Articles 4 and 12 of the UNFCCC, China submitted National Communications dated 2004, 2012, and 2018 and Biennial Update Reports dated 2016 and 2018.[110]

105 It covers impacts and challenges of climate change in China; China's corresponding efforts, guidelines, principles, objectives, and measures in relation to climate change and China's positions on key climate change issues; the need for international cooperation. See China's National Climate Change Programme (中国应对气候变化国家方案) State Council Order No. (2007)17, 3 June 2007, section 1 to section 5.

106 Wang (n 5), 111. Pan (n 5), 80.

107 China's National Climate Change Programme (n 105), part three.

108 Ibid.

109 Pan (n 5), 80. Department of Climate Change, National Development and Reform Commission of People's Republic of China, *Enhanced Actions on Climate Change: China's Intended Nationally Determined Contributions* (2015) Beijing, 30 June 2015, 3. See also UNFCCC website, <http://unfccc.int/resource/climateaction2020/>, where it is stated:

> China stated that the above-mentioned autonomous domestic mitigation actions are voluntary in nature and that they will be implemented in accordance with the principles and provisions of the Convention, in particular Article 4, paragraph 7. The Party also stated that its communication is made in accordance with the provisions of Article 12, paragraphs 1(b) and 4, and Article 10, paragraph 2(a), of the Convention.

110 'National Communication Submissions from Non-Annex I Parties', <https://unfccc.int/process-and-meetings/transparency-and-reporting/reporting-and-review-under-the-convention/national-communications-and-biennial-update-reports-non-annex-i-parties/national-communication-submissions-from-non-annex-i-parties> accessed 29 November 2019. Biennial Update Report Submissions from Non-Annex I Parties, <https://unfccc.int/BURs> accessed 29 November 2019. For the latest national communication and the biennial report, see Ministry of Ecological Environment, The People's Republic of China Third National Communication on Climate Change (中华人民共和国气候变化第三次国家信息通报), submitted to the UNFCCC secretariat in June 2019, <https://unfccc.int/documents/197660> accessed 29 November 2019; and Ministry of Ecological Environment, The People's Republic of China Second Biennial Update Report on Climate Change (中华人民共和国气候变化第二次两年更新报告), submitted to the UNFCCC secretariat in June 2019, <https://unfccc.int/documents/197666> accessed 29 November 2019.

The Kyoto Protocol

The Kyoto Protocol was adopted in December 1997 and entered into force on 16 February 2005. It has 192 parties. China signed it on 29 May 1998 and ratified it on 30 August 2002.

Article 12 of the Kyoto Protocol established the Clean Development Mechanism (CDM)[111] to encourage participation by developing-country parties, including China, in the implementation of the Kyoto Protocol.[112] In fact, China has been the leading participant in CDM projects. To regulate such projects, China adopted the Administrative Measures for the Operation and Management of Clean Development Mechanism Projects.[113] The Administrative Measures provide for institutional structure, application, implementation procedures, and liability rules for CDM projects in China.

The Paris Agreement

The Paris Agreement was adopted on 12 December 2015 and entered into force on 4 November 2016. China signed it on 22 April 2016 and ratified it on 3 September 2016. The targets set by the Paris Agreement include the following:

(a) Holding the increase in the global average temperature to well below 2°C above pre-industrial levels and pursuing efforts to limit the temperature increase to 1.5°C above pre-industrial levels, recognizing that this would significantly reduce the risks and impacts of climate change.
(b) Increasing the ability to adapt to the adverse impacts of climate change and foster climate resilience and low greenhouse gas emissions development, in a manner that does not threaten food production.
(c) Making finance flows consistent with a pathway towards low greenhouse gas emissions and climate-resilient development.[114]

The Paris Agreement sets a higher standard in certain matters for developed-country as opposed to developing-country parties,[115] in accordance with the principle of common but differentiated responsibilities and respective capabilities.[116]

111 Javier De Cendra De Larragán, 'The Kyoto Protocol, with a Special Focus on the Flexible Mechanisms', in Daniel A. Farber and Marjan Peeters (eds), *Climate Change Law,* 229 (Elgar Encyclopedia of Environmental Law, Edward Elgar, 2016); Xiaoyi Jiang, *Legal Issues for Implementing the Clean Development Mechanism in China,* 19–20 (Springer, 2013).
112 Larragán (n 111), 230.
113 Administrative Measures for the Operation and Management of Clean Development Mechanism Projects (清洁发展机制项目运行管理办法), National Development and Reform Commission, the Ministry of Science and Technology, Ministry of Foreign Affairs, and Ministry of Finance; adopted in 2005, amended by Order No. (2011)11, 3 August 2011, art. 1.
114 Paris Agreement, 12 December 2015, art. 2, para. 1.
115 Ibid., art. 4, para. 4.
116 Ibid., art. 2, para. 2.

However, the distinction between Annex I and non-Annex I parties has been removed, suggesting a greater continuity of legal responsibility.

Nationally determined contributions (NDCs) are the new method for parties to make commitments on their post-2020 climate actions.[117] Article 3 of the Paris Agreement requires all parties to prepare and communicate NDCs.[118] NDCs are to facilitate the achievement of the targets of the Paris Agreement.[119]

China communicated its intended nationally determined contribution (INDC) in June 2015 and submitted its first NDC (which was, in fact, the INDC unchanged) to the interim NDC registry in September 2016. In addition to adaptation actions, China pledged to achieve the following goals by 2030: to peak CO_2 emissions around 2030 and to make best efforts to peak early; to lower CO_2 emissions per unit of GDP by 60%–65% from the 2005 level; to increase the share of non-fossil fuels in primary energy consumption to around 20%; and to increase the forest-stock volume by around 4.5 billion cubic metres above the 2005 level.[120]

NDCs are to be updated every five years.[121] The Conference of the Parties to the Paris Agreement is to undertake a global stocktake every five years, with the first taking place in 2023.[122] NDCs must become more ambitious over time and are not allowed to backtrack on their ambition.[123]

Article 13 of the Paris Agreement established an enhanced transparency framework (ETF), setting generally uniform requirements but also some different requirements for developed-country parties on reporting, technical expert review, and multilateral consideration of progress.[124] Developing-country parties are to be given financial and other support for the effective implementation of the Paris Agreement.[125]

Chinese laws and policies related to climate change

Given that there is no indication that international law related to climate change is self-executing, it must first be ratified by the NPC or its standing committee and then incorporated into domestic law.[126] Given the targets announced by China in 2007, the country can be said to have taken measures to address climate change

117 Nationally Determined Contributions (NDCs), <https://unfccc.int/process-and-meetings/the-paris-agreement/nationally-determined-contributions-ndcs#eq-2> accessed 29 November 2019.
118 Paris Agreement (n 114), arts. 3 and 4, para. 2.
119 Ibid. Nationally Determined Contributions (NDCs) (n 117).
120 Ministry of Ecological Environment, 3rd NC (n 110).
121 Paris Agreement (n 114), art. 4, para. 9.
122 Ibid., art. 14, para. 2.
123 Ibid., arts. 3 and 4, para. 3.
124 Reporting and Review under the Paris Agreement, <https://unfccc.int/process-and-meetings/transparency-and-reporting/reporting-and-review-under-the-paris-agreement/reporting-and-review-under-the-paris-agreement> accessed 29 November 2019. For a more detailed discussion, see Chapter 7 of this book.
125 Paris Agreement (n 114), art. 3.
126 Guo (n 54), 45–46.

since at least that date.[127] However, from a legal perspective, it is important to understand whether and how the requirements of international obligations have been prescribed in laws and policies in China. Several laws (*fǎlǜ*), regulations, and administrative measures deal with climate change, through topics such as environmental protection, meteorological law, intellectual property law, low-carbon cities, emission-trading systems, energy law, and natural resources law.[128] Policies are important in the climate law system in China, and they will be analysed separately.

Constitution

The Constitution of China provides an institutional foundation for many policy fields, including the regulation of climate change.[129] Articles 2 and 26 of the Constitution are applicable to the topic of climate change. In accordance with Article 26, the state's responsibility is to protect and improve the environment and to prevent and control pollution and other public hazards. The state also has the responsibility to organize and encourage afforestation and the protection of forests. Because climate change is an important environmental issue and afforestation is related to climate change mitigation and adaptation, this provision provides a legal basis for adopting laws and policies to address climate change in China.[130]

Article 2(3) of the Constitution is a basis for public involvement in environmental protection.[131] In accordance with this provision, the public is to be involved through various channels in the management of three types of issues: state affairs, economic and cultural undertakings, and social affairs. The government is to protect this constitutional right of the public.[132] Accordingly, public involvement is to be allowed in the context of climate change, as the latter is an aspect of "social affairs".[133] However, rules need to be adopted to support the right of the public to participate in climate-related actions. This can be achieved by either the Constitution or a future climate change law embodying substantive environmental rights.[134]

127 Wang and Tang (n 3), 120.

128 Liu Zhe, 'The Basic Problems in China's Legislation on Climate Change' (我国应对气候变化立法的基本问题探究) (2016) 18 *Journal of Jiangsu University (Social Science Edition)* (江苏大学学报(社会科学版)) 16–18.

129 Zhou (n 20), 171.

130 Guo Dongmei, *Studies on the Legal System in Relation to Climate Change* (应对气候变化法律制度研究), 145 (Law Press, 2010).

131 Wang (n 48), 30.

132 It is a constitutional right of the public to get involved in the three types of issues. See Ji Yaping, 'Studies on Public Participation in Administrative Decision-Making' (行政决策程序中的公众参与研究) (2012) 8 *Zhejiang Academic Journal* (浙江学刊) 166.

133 Huanling Lang, *Public Participation in Environmental Decision-Making in China: Towards an Ecosystem Approach* (Doctoral Dissertation, University of Groningen, 2014) 140; Wang (n 48), 30.

134 Zhou Xian and Chen Demin, 'Rethinking and Reconstructing Public Participation in Climate Change Response' (公众参与气候变化应对的反思与重塑) (2019) 29 *China Population, Resources and Environment* (中国人口·资源与环境) 117.

Laws (fǎlǜ)

As mentioned, several of China's laws (*fǎlǜ*) are related to climate change.[135] These cases come in three types: laws adopted with the aim of environmental protection; energy law; and laws that contain adaptation measures.[136] Many environmental protection laws in China apply to climate change issues. Moreover, many energy laws promote the aim of reducing GHG emissions. De facto adaptation measures exist in the Chinese legal system, adopted not for the purpose of addressing climate change but nevertheless helping to do so. There are overlapping objectives among the three categories. For instance, some laws deal with environmental protection and energy conservation, and some deal with environmental protection and adaptation to climate change.[137]

Environmental protection laws (fǎlǜ)

According to Article 2 of the Environmental Protection Law (EPL), *environment* includes the atmosphere, grasslands, wildlife, and forests. The changed composition of the atmosphere leads to climate change, while grasslands and forests are carbon sinks. Moreover, climate change has impacts on ecosystems that are protected by the EPL.[138] Therefore, although climate change is not explicitly regulated by the EPL, this law can be used in response to climate change. In particular, Article 6 of the EPL requires citizens, enterprises, and governments at all levels to protect the environment, and citizens are encouraged to lead a low-carbon lifestyle. Article 22 enables the use of taxation to control the discharge of pollutants.[139] Articles 53–58 protect the rights of the public on access to environmental information and participation in decision-making on environmental protection.

135 Pan (n 5), 81.

136 As mentioned earlier, laws such as meteorological law, intellectual property law, and natural resources law are also related to climate change. Some laws fit into one of the three types. For example, natural resources law is linked to climate adaptation. Some laws, such as intellectual property law, provide important tools for achieving goals of climate change activities.

137 For instance, the Renewable Energy Law aims to promote the development and use of renewable energy, and it also encourages environmental protection. See Renewable Energy Law of the People's Republic of China (中华人民共和国可再生能源法), Standing Committee of the National People's Congress (adopted on 28 February 2005, amended on 26 December 2009), art. 1. In addition to environmental protection, the Environmental Protection Law also has effects on energy conservation and climate change adaptation.

138 Zhou (n 20), 173; Frank Maes, An Cliquet, Willemien du Plessis, and Heather McLeod-Kilmurray, *Biodiversity and Climate Change* (Edward Elgar, 2013), Introduction, xvi–xvii; The Environmental Protection Law of the People's Republic of China (中华人民共和国环境保护法), Standing Committee of the National People's Congress, adopted on 26 December 1989, amended on 24 April 2014, art. 30.

139 Environmental Protection Tax Law regulates the taxation of pollutants. However, GHGs are not included as pollutants. See Environmental Protection Tax Law of the People's Republic of China (2018 Revision) (中华人民共和国环境保护税法(2018修正)), Standing Committee of the National People's Congress, 26 October 2018, arts. 1 and 3. One suggestion in the literature is to enlarge the environmental protection tax to include a carbon tax. See Wang Guofei, 'Obstacle and Breakthrough: A Carbon Tax Included in the Environmental Protection Tax Law' (障碍与突围:碳税纳入环境保护税的思考) (2015) 43 *Environmental Protection* (环境保护) 37–38.

Articles 31–32 establish compensation and remediation systems for the environment, which are a kind of climate adaptation measure. Article 40 promotes cleaner production and resource recycling by imposing obligations on governments and enterprises – contributing to the reduction of GHG emissions.

The Law on Prevention and Control of Atmospheric Pollution has as one of its aims to promote coordinated control over air pollutants and GHGs.[140] Reference to GHGs provides an opportunity to integrate climate change considerations into the Law on Prevention and Control of Atmospheric Pollution.[141] However, the law neither mentions GHGs in other provisions nor elaborates on how air pollutants and GHGs are to be controlled.[142] While the law does not clearly define "air pollutant", it lists the main air pollutants: particles, sulphur dioxide, nitrogen oxides, volatile organic compounds, and ammonia.[143] Since no GHG is listed by the law as an air pollutant, climate change issues are not regarded as air-quality issues.[144] The Kyoto Protocol defines six GHGs, consisting of carbon dioxide, methane, nitrous oxide, hydrofluorocarbons, perfluorocarbons, and sulphur hexafluoride.[145] As argued by some scholars, some GHGs, such as nitrous oxide, have features of air pollutants and could therefore be controlled by the Law on Prevention and Control of Atmospheric Pollution.[146]

The Circular Economy Promotion Law[147] and the Cleaner Production Promotion Law[148] also contribute to the control of GHG emissions, since they encour-

140 Atmospheric Pollution Prevention and Control Law of the People's Republic of China (中华人民共和国大气污染防治法), Standing Committee of the National People's Congress (adopted on 5 September 1987, amended on 29 August 1995 and 26 October 2018), art. 2, para. 2.

141 Yamineva and Liu (n 5), 4.

142 Ibid. Liu Jing, 'The Legality for Greenhouse Gas Emission Reduction: A Coordinated Control of Greenhouse Gases and Air Pollutants – a Comment on Article 2, Paragraph 2, of the Law on Prevention and Control of Atmospheric Pollution' (温室气体减排的法律路径：温室气体和大气污染物协同控制 – 评《大气污染防治法》第2条第2款) (2019) 47 *Journal of Xinjiang University (Philosophy, Humanities & Social Science)* (新疆大学学报(哲学·人文社会科学版)) 52; Zhang Qijing, 'Research on Response of Climate Change Legislation in Post-Paris Era' (后巴黎时代中国应对气候变化立法研究) (2017) 29 *Environment and Development* (环境与发展) 2.

143 Atmospheric Pollution Prevention and Control Law (n 140), art. 2, para. 2.

144 Yamineva and Liu (n 5), 7. Because a climate change law has not been adopted, it is still not clear how GHGs will be defined in the Chinese legal system.

145 Kyoto Protocol to the United Nations Framework Convention on Climate Change, United Nations Treaty Series, vol. 2303, p. 162, Kyoto, 11 December 1997, Annex A.

146 Liu (n 142), 51; Yamineva and Liu (n 5), 8; Zhou Xiaoguang and Zhang Jianwei, 'Thoughts on the Coordinated Governance of Air Pollution and Climate Change' (关于大气污染与气候变化协同治理的法律思考) (2017) 5 *Tribune of Social Sciences* (社会科学论坛) 222.

147 It is formulated to promote the development of the circular economy, improving the resource use efficiency, protecting and improving the environment, and realizing sustainable development. See Circular Economy Promotion Law of the People's Republic of China (中华人民共和国循环经济促进法), Standing Committee of the National People's Congress (adopted on 29 August 2008, amended on 26 October 2018), art. 1. See also, Zhou (n 20), 186–188.

148 It aims to promote cleaner production, raise the efficiency of use of resources, reduce and avoid the generation of pollutants, protect and improve environment, protect human health, and promote the sustainable development of economy and society. See Cleaner Production Promotion Law of the People's Republic of China (中华人民共和国清洁生产促进法), Standing Committee of the National People's Congress (adopted on 29 June 2002, amended on 29 February 2012), art. 1. See also, Zhou (n 20), 188–189.

age the development of a low-carbon society and take measures to promote the efficient use of resources. They can contribute to GHG-emission reduction from industry.[149]

*Energy conservation laws (*fǎlǜ*)*

China has enacted several energy-related laws, such as the Renewable Energy Law,[150] the Energy Conservation Law,[151] the Electric Power Law,[152] and the Coal Industry Law.[153] They encourage the use of renewable and clean-energy resources and emphasize the protection of the environment during the use of resources.[154] Although they were not designed to combat climate change, climate change mitigation can be achieved if they are complied with, so their role is important in China.[155]

*Laws related to climate adaptation (*fǎlǜ*)*

The Meteorology Law aims, among other things, to protect climatic resources.[156] Chapter 4 of the law sets requirements on meteorological forecasts and severe weather warnings. Since climate change can cause abnormal weather patterns, rules under this law can help to address the adverse effects of climate change. Other laws, such as the Forest Law, the Grassland Law, and the Water and Soil

149 The relevant provisions are still abstract. See Pan (n 5), 81.

150 It aims to promote the development and use of renewable energy, increase the supply of energy, improve the structure of energy, safeguard energy security, protect the environment, and realize sustainable economic and social development. See Renewable Energy Law (n 137), art. 1.

151 It aims to promote energy conservation in the whole society, enhance energy use efficiency, protect and improve the environment, and promote comprehensive, coordinated, and sustainable economic and social development. See Energy Conservation Law of the People's Republic of China (中华人民共和国节约能源法), Standing Committee of the National People's Congress (adopted on 1 November 1997, amended on 2 July 2016 and 26 October 2018), art. 1.

152 It is enacted to guarantee and promote the development of the electric power industry; to safeguard the lawful rights and interests of those who invest in, manage, or consume electric power; and to guarantee the safe operation of electric power. See Electric Power Law of the People's Republic of China (中华人民共和国电力法), Standing Committee of the National People's Congress (adopted on 28 December 1995, amended on 27 August 2009, 24 April 2015, and 29 December 2018), art. 1.

153 It is enacted to develop, utilize, and protect the coal resources, standardize the production and marketing of coal, and promote and ensure the development of the coal industry. See Coal Industry Law of the People's Republic of China (中华人民共和国煤炭法), Standing Committee of the National People's Congress (adopted on 29 August 1996, amended on 27 August 2009, 22 April 2011, 29 June 2013, and 7 November 2016), art. 1.

154 Energy Conservation Law (2018 Amendment) (n 151), art. 7, para. 3; Renewable Energy Law (n 137), art. 4; Electric Power Law (n 152), art. 5; Coal Industry Law (n 153), art. 11.

155 Zhou (n 20), 180.

156 Meteorology Law of the People's Republic of China (中华人民共和国气象法), Standing Committee of the National People's Congress (adopted on 31 October 1999, amended on 27 August 2009, 31 August 2014, and 7 November 2016), art. 1.

Conservation Law, are relevant to climate change in the sense that they contribute to climate adaptation.[157] As explained at the beginning of this subsection, some laws have overlapping objectives. The foregoing laws are related to both environmental protection and adaptation. They may be categorized as laws relating to climate adaptation because they are more about adaptation measures when addressing climate change.

Regulations

The Regulation on Energy Conservation Management for Public Buildings and the Regulation on Energy Conservation for Civil Buildings were enacted by the State Council to enforce the Energy Conservation Law.[158]

Some local congresses and governments have enacted regulations on climate change. Qinghai province was the first to adopt a local regulation, the Regulation Addressing Climate Change,[159] in 2010; it was followed by Shanxi province.[160] In both regulations, adaptation and mitigation are equally emphasized. However, their operational provisions and liability rules still remain to be developed.[161]

Administrative measures

As noted earlier, the Administrative Measures for the Operation and Management of Clean Development Mechanism Projects were developed by the NDRC, the Ministry of Science and Technology, the Ministry of Foreign Affairs, and the Ministry of Finance in 2011.[162] In 2014, the NDRC released the Interim Measures

157 Forest Law of the People's Republic of China (中华人民共和国森林法), Standing Committee of the National People's Congress (adopted on 20 September 1984, amended on 27 August 2009); Water and Soil Conservation Law of the People's Republic of China (中华人民共和国水土保持法), Standing Committee of the National People's Congress (adopted on 29 June 1991, amended on 25 December 2010); Grassland Law of the People's Republic of China (中华人民共和国草原法), Standing Committee of the National People's Congress (adopted on 18 June 1985, amended on 27 August 2009 and 29 June 2013).

158 Regulation on Energy Conservation Management for Public Buildings (公共机构节能条例) State Council Order No. (2008)531, 1 August 2008, amended on 1 March 2017; Regulation on Energy Conservation for Civil Buildings (民用建筑节能条例), State Council Order No. (2008)530, 23 July 2008.

159 Regulation on Addressing Climate Change of Qinghai Province (青海省应对气候变化办法), People's Government of Qinghai Province (adopted on 23 July 2010).

160 Regulation on Addressing Climate Change of Shanxi Province (山西省应对气候变化办法), People's Government of Shanxi Province (adopted on 12 July 2011); Zhang (n 142), 2.

161 Tian Danyu, 'Studies on Local Laws on Climate Change' (中国地方应对气候变化立法研究) (2018) 3 *Law-Based Society* (法治社会) 80–81; Peng Benli and He Yan, 'Legislation at the Local Level to Address Climate Change in China: Features, Limitations and Prospects' (中国应对气候变化地方立法 – 特点、局限与展望) (2017) 38 *Journal of Yulin Normal University (Philosophy & Social Science)* (玉林师范学院学报(哲学社会科学)) 67–68.

162 Administrative Measures for the Operation and Management of Clean Development Mechanism Projects (n 113).

for the Administration of Carbon Emission Permit Trading.[163] Because the Interim Regulation (not measure) for the Administration of Carbon Emission Permit Trading is still being developed, the two administrative measures guide emission-trading projects.[164]

Policies

In 2008 and 2011, China published White Papers on climate change governance.[165] Since 2009, the country has also published annual reports on its policies and actions on climate change.[166] The NDRC published the Notice on Launching Carbon Emissions Trading Pilots in 2011 to guide ETS pilot projects (of which there were seven, in total).[167] The National Strategy of Climate Change Adaptation was released in 2013, explaining Chinese plans on climate change adaptation.[168] The Thirteenth Five-Year Plan specifies climate change policies and actions in its Chapter 46.[169] Adaptation and mitigation are two main aspects, and international cooperation is encouraged.[170] The State Council adopted the National Plan on Climate Change in 2014 and further adopted the 13th Five-Year Work Plan for Controlling Greenhouse Gas Emission in 2016.[171]

163 Interim Measures for the Administration of Carbon Emission Permit Trading (碳排放权交易管理暂行办法), National Development and Reform Commission Order No. (2014)17, 10 December 2014.

164 The Interim Regulation for the Administration of Carbon Emission Permit Trading (碳排放权交易管理暂行条例) was drafted by MEE. MEE solicited public opinions on this interim regulation from March to May 2019. See SCIO Briefing on China's Policies and Actions for Addressing Climate Change, <http://english.scio.gov.cn/pressroom/2019-11/28/content_75458089.htm> accessed 29 November 2019.

165 China's Policies and Actions on Climate Change (2011) (中国应对气候变化的政策与行动(2011)), Information Office of the State Council of the People's Republic of China, <www.china.com.cn/ch-book/node_7142682.htm> accessed 29 November 2019.

166 They are usually released before the opening of COPs of the UNFCCC. The latest annual report was released on 27 November 2019, see SCIO Briefing on China's Policies and Actions for Addressing Climate Change (n 164).

167 Notice of the General Office of the NDRC on Launching Carbon Emissions Trading Pilots (国家发展改革委关于开展碳排放权交易试点工作的通知), General Office of the NDRC Order No. (2011)2601, 29 October 2011. The national carbon emission trading market was launched in 2017. See Ministry of Ecological Environment, 3rd NC (n 110). For a more detailed discussion on policies in relation to ETS, see Chapter 5 of this book.

168 National Strategy of Climate Change Adaptation (国家适应气候变化战略), <www.gov.cn/gzdt/att/att/site1/20131209/001e3741a2cc140f6a8701.pdf> accessed 29 November 2019. For more on adaptation, see Chapter 8 of this book.

169 The Thirteenth Five-Year Plan for National Economic and Social Development of the People's Republic of China (n 91), chapter 46.

170 Ibid.

171 NDRC, National Plan on Climate Change (2014–2020) (n 19), 'The 13th Five-Year Work Plan for Controlling Greenhouse Gas Emission' <www.gov.cn/zhengce/content/2016-11/04/content_5128619.htm> accessed 29 November 2019.

Chinese institutional settings related to climate change

In China, many administrative authorities are involved in the regulation of climate change. Both the central government and local governments are competent in dealing with climate change. Within the central government, the National Leading Group to Address Climate Change and Energy Conservation is an inter-ministerial group that coordinates 30 ministries and commissions that are involved in climate change and energy conservation.[172] The State Council is in charge of the Leading Group.[173] One responsibility of the Leading Group was to fulfil China's obligations under the UNFCCC and the Kyoto Protocol.[174] The Leading Group also develops national strategies and policies on climate change, makes overall arrangements on activities that address climate change, and arranges matters in relation to energy conservation and emission reduction.[175] China's NDC was prepared by the Leading Group.[176] The Ministry of Ecology and Environment (MEE) and the NDRC are responsible for the daily work of the Leading Group.[177] Other ministries or commissions appoint responsible units to address climate change and guide industries under their supervision.[178]

The NDRC was the central government body for climate change before 2018. The Department of Climate Change was established within the NDRC in 2008. The National Center for Climate Change Strategy and International Cooperation (NCSC) was established in 2012. It aims to undertake research on climate change issues and provide technical support for developing laws and policies related to climate change, implementing emission-trading systems, and supporting international negotiations on climate change.[179] After the institutional reform of 2018, some of the NDRC's responsibilities on climate change were transferred to the

172 Notice of the State Council Regarding the Establishment of the National Leading Group to Address Climate Change and Energy Conservation (国务院关于成立国家应对气候变化及节能减排工作领导小组的通知), State Council Order No. (2007)18, 12 June 2007; Notice of the General Office of the State Council Regarding the Changes of Members of the National Leading Group to Address Climate Change and Energy Conservation (国务院办公厅关于调整国家应对气候变化及节能减排工作领导小组组成人员的通知 [2019]), General Office of the State Council Order No. (2019)99, 7 October 2019.

173 The premier is the head of the Leading Group.

174 Tian Danyu, 'An Evaluation and Analysis on the Climate Governance Organization Nowadays' (中国目前气候治理组织机构评析) (2013) 1 *Journal of China University of Political Science and Law* (中国政法大学学报) 147.

175 Notice of the State Council Regarding the Establishment of the National Leading Group to Address Climate Change and Energy Conservation (n 172).

176 Ministry of Ecological Environment, 2nd BUR (n 110), 5.

177 This is consistent with the changes in institutional arrangements regarding climate change. Before 2018, the NDRC was the major commission that was responsible for the daily work of the Leading Group.

178 Ministry of Ecological Environment, 2nd BUR (n 110), 6.

179 Objectives and Missions (中心职责). <www.ncsc.org.cn/gyzx/zxzer/> accessed 29 November 2019.

MEE.[180] The NDRC still works on climate change issues because, first, it is still responsible for energy conservation and because, second, it remains one of the two leading departments in the National Leading Group to Address Climate Change and Energy Conservation.[181] The NCSC is now directed by the MEE rather than the NDRC. Meanwhile, the Department of Climate Change was transferred from the NDRC to the MEE to lead climate change actions and the reduction of GHG emissions. That department is also responsible for designing and implementing policies and programmes on climate change, taking actions to implement China's obligations under the UNFCCC, and cooperating with other government bodies to participate in international negotiations and conferences.[182] The MEE is to promote coordinated control over climate change and air pollution. However, the MEE first has to have a specific legal basis to deal with climate change issues, because, as explained earlier, measures are not provided for controlling GHGs in the Law on Prevention and Control of Atmospheric Pollution.[183]

At the local level, leading groups on climate change are established by the provincial governments.[184] Provincial governors chair leading groups on climate change in provincial governments.[185] These leading groups address climate change issues by coordinating departments, such as those on environmental protection, natural resources, science and technology, and finance.[186] Consistent

180 State Council Institutional Reform Plan of 2018 (n 44). According to Zhao Yingmin, vice minister of MEE, the institutional adjustment regarding climate change was completed by the end of November 2019. See SCIO Briefing on China's Policies and Actions for Addressing Climate Change (n 164).

181 The Department of Resource Conservation and Environmental Protection within the NDRC undertakes concrete tasks in relation to the Leading Group. The NDRC's responsibilities focus on energy conservation. See Allocation of Functions and Responsibilities and Internal Departments (职能配置与内设机构), <www.ndrc.gov.cn/fzggw/bnpz/> accessed 6 December 2019. This is consistent with the fact that the NDRC still oversees the National Energy Administration.

182 'Department of Climate Change', <http://english.mee.gov.cn/About_MEE/Internal_Departments/201605/t20160526_346894.shtml> accessed 29 November 2019.

183 For example, measures need to be taken to build the personnel capacity regarding climate change and integrate climate change measures into the already-existing environmental governance system. See Zhai Qing, 'Pushing the Modernization of China's Governance System and Capacity of Environmental Governance to Win the Tough Fight against Environmental Pollution' (推进生态环境治理体系和治理能力现代化 为打好污染防治攻坚战提供坚强保障) (2018) 10 *China Institutional Reform and Management* (中国机构改革与管理) 14.

184 Ministry of Ecological Environment, 2nd BUR (n 110), 6–7; Schreurs (n 7), 169.

185 Ministry of Ecological Environment, 2nd BUR (n 110), 7.

186 Ibid. To take the leading group in Hubei province as an example, the governor of Hubei province chairs the leading group, and the leading group coordinates at least 25 departments in the provincial government. Similar to the arrangement in the National Leading Group, the Department of Ecological Environment and Hubei Provincial Development and Reform Commission are responsible for the daily work of the provincial leading group within their competences respectively. See Notice of the General Office of Hubei province on the Adjustment of the Names and Members of the Leading Group on Energy Conservation (Climate Change) (省人民政府办公厅关于湖北省节能减排（应对气候变化）工作领导小组更名并调整组成人员的通知), General office of Hubei province, document number 22, 20 March 2019.

with the institutional arrangements at the central government level, the provincial governments also establish divisions that address climate change issues within departments that are responsible for environmental protection.[187]

Conclusion

Parties to the UNFCCC and the Paris Agreement are required to take measures to address climate change, but they are not obliged to adopt national laws to do so. Policies that address climate change are therefore one way for China to fulfil its international obligations on climate change. In fact, both laws and policies are being used by China to deal with climate change issues.

Although several laws (*fǎlǜ*), regulations, and administrative measures in China can be implemented to address climate change issues, none of them was enacted to address climate change as such. The lack of a clear statement of an objective to combat climate change constrains the applicability of some legal measures. A systematic legal framework for climate change is absent. For example, adaptation measures are not adequately provided for in the current legal system. A climate change law with a clear aim to combat climate change that covers all necessary mitigation and adaptation measures is necessary to improve China's capacity to deal with climate change. In the meantime, China is taking the practical approach of first adopting regulations for specific issues, such as emission-trading systems, before going further. This approach may finally lead to the adoption of a climate change law, after sufficient experience with the regulation of climate change has been gained.

187 For example, the Division of Climate Change within the Department of Ecological Environment of Shandong province and the Division on Science and International Cooperation (Division of Climate Change) within the Department of Ecological Environmental of the government of Beijing municipality are responsible for addressing climate change issues in each of the two provincial governments: <http://sthj.shandong.gov.cn/jggk/jgcs/> and <http://sthjj.beijing.gov.cn/bjhrb/xxgk/jgzn/jgsz/jjgjgszjzz/index.html> accessed 6 December 2019. Provincial governments have incentives to address climate change issues and have contributed to the reduction of GHG emissions by, for instance, initiating pilot regional emission-trading schemes. See Schreurs (n 7), 169–171.

2 Scholarship and running debates on climate law and policy in China

Xiangbai He

Introduction

Scholarly discussion on climate law and policy has formed an integral part of China's legal scholarship. Understanding and analysing this scholarship is important because it not only reveals the key legal concerns related to climate change among Chinese scholars but also signals China's future regulatory path towards improved climate change governance.

As a latecomer to the climate change arena, China's scholarly discussion also started late. A general observation on the Chinese literature is that from 2008 onwards there has been a significant increase in legal scholarship on climate change, partly due to the commencement of the first commitment period of the Kyoto Protocol (KP). The fact that the Chinese government has switched to a proactive stance on climate governance and low-carbon development since 2008 also has contributed to the boom in climate change literature in China.

The early stage of climate law scholarship was preoccupied with how China should respond to climate change treaties, in particular the KP; how China could reform its legal system to respond to climate change; and how China could learn from other developed countries, such as the United States (US), United Kingdom (UK), Japan, and Germany. Recent climate law scholarship in China has been diverse, including a focus on climate change legislation, other legal instruments for regulating climate change, sector-based climate change responses to climate change liability, and even climate litigation. This evolution reveals that the research focus has been shifting from prescriptive studies to more empirical ones. Additionally, the conceptual understanding of climate change has changed from being a problem, or challenge, that needs to be addressed to being a background against which China's legal system and regulatory instruments should be reviewed and developed.

Nonetheless, climate law scholarship in China has lagged behind the mainstream scholarship on environmental law. This is apparent from a survey of climate law articles published in top Chinese law journals and written by well-known environmental law scholars. A key reason for the lag is that among Chinese legal scholars, environmental pollution issues are regarded as more important than climate change issues for China. Limited by this perception, scholarly legal research on climate change in China is comparatively incomplete, unsystematic, and immature. Yet it is still possible to profile China's scholarly contribution to climate law, to understand how climate change has been defined, framed, and analysed by legal scholars in the Chinese-language literature.

Although this chapter focuses on the climate law literature in Chinese, there have been several works on China's climate law written in English. Due to the language barrier, works in Chinese have received little attention outside China. This chapter attempts to make up for this gap. Using the China National Knowledge Internet (CNKI) as the main source of information, the chapter finds that China's literature on climate law focuses on four topics in general:

1. Is a stand-alone climate change law for China essential and feasible? If yes, how should it be formulated?
2. Should CO_2 be regulated as an ordinary pollutant through air-pollution law?
3. Which regulatory instrument, including carbon trading or a carbon tax, would work best to mitigate GHG emissions in the Chinese context?
4. What are the key legal issues for a carbon-trading scheme in China?

Key legal debates on climate change in China

Whether to make, and how to make, a stand-alone climate change law

In 2009, the National People's Congress (NPC) Standing Committee adopted a *Resolution of the Standing Committee of the National People's Congress on Making Active Responses to Climate Change.*[1] As the first resolution adopted by the top legislative body in China to address climate change, it proposes to strengthen China's legal framework in addressing climate change and states that climate-change-related legislation will be incorporated into future legislative agendas. However, it does not clarify whether a stand-alone climate law is necessary. In 2011, both the National Development and Reform Commission (NRDC) and the Ministry of Environmental Protection (MEP, the predecessor of the Ministry of Ecology and Environment) initiated research programmes to draft their own versions of a climate law. The Chinese Academy of Social Sciences (CASS), a think tank of the State Council, under request by the MEP, published a draft text in March 2012.[2] The regulatory scope of this draft is comprehensive and includes general principles and specific mechanisms to address climate change. According

1 *Resolution of the Standing Committee of the National People's Congress on Making Active Responses to Climate Change*, People's Republic of China, Standing Committee of National People's Congress (全国人大常委会关于积极应对气候变化的决议) (2 August 2009).

2 Chinese Academy of Social Sciences, Statement on the Law on Addressing Climate Change (Proposal) (关于《气候变化应对法》(建议稿)的说明), 2012.

to the draft, responsibility for addressing climate change would be distributed among different related authorities, legal institutions to address the mitigation of climate change would be established (such as a total volume control mechanism, eco-compensation, a low-carbon criterion, an emission-trading scheme, and a carbon tax), and supporting measures would be provided. However, since the draft text was not adopted by the NPC or its standing committee, scholarly opinions are still divided on whether a stand-alone climate law is needed and how this law should be formulated.

There was some support for integrating climate change issues into the regulatory framework being developed for China's Energy Law[3] and into China's Law on Prevention and Control of Atmospheric Pollution (LPCAP),[4] yet the idea of making a specialized climate law gained the support of most scholars, who justified their support by reference to the failure of existing laws and regulations, as well as the entrenched institutional structure, to deal with the complexity and uncertainty of climate change.[5]

Tian's argument in favour of climate change legislation is representative: it is necessary to provide a legal status for targets on carbon-emission peaking, total carbon emissions, and carbon-intensity control; clarify the legal nature of carbon-emission rights; and provide a legal basis for distributing and enforcing emission-reduction targets.[6] Tian also argues that making a specialized law is feasible in China because the country has gained sufficient legislative experience on carbon reduction, local-level climate change legislation, and information from other countries.[7] Nevertheless, the legal issues and challenges of making a specialized climate law have not been fully exposed or appreciated in the literature.

The first issue discussed by legal scholars is the legislative model. For China, the question arises whether a framework legislation on climate change or a comprehensive law would work better, given its existing laws and institutions. A framework law would lay down general principles and institutional arrangements for climate change, the implementation of which would need to be supplemented by

3 Energy Research Institute, China Law Society, *China Energy Law Research Report* (2009) (中国能源法研究报告(2009)) (Law Press China, 2010). The Energy Law has not yet been promulgated.

4 Meng Wei, 'Integrating GHGs Emission in the Law on Prevention and Control of Atmospheric Pollution' (孟伟：将温室气体排放问题纳入《大气污染防治法》) (Ministry of Natural Resources, 11 March 2010) <www.mnr.gov.cn/zt/hu/lh/2010qglh/lhhzy/gzhj/201003/t20100313_2022743.html> (accessed 31 October 2019).

5 Qin Dahe, 'Developing the Legal System to Respond to Climate Change' (完善法律制度，应对气候变化) *Guangming Daily* (20 April 2009); Zhou Hongchun, 'The Practical Significance and Policy Suggestions of Developing a Low-carbon Economy in China' (中国发展低碳经济的现实意义及政策建议) *China Economic Times*, 28 December 2009.

6 Tian Danyu, 'Research on Climate Change Legislation' (应对气候变化立法研究) (2018) 3 *World Environment* (世界环境) 59–62.

7 Ibid.

lower-level legislation or sector-based regulation.[8] Comprehensive legislation, by contrast, would not only specify principles and establish institutions but would also clarify detailed regulatory goals, mitigation and adaptation measures, and regulatory instruments.[9] As argued by Wang and Liu, for any country, the selection of the legislation model depends on the legislative tradition, the intended content of the legislation, and the legislative background in related regulated areas.[10] For example, the United Kingdom and Mexico have each adopted comprehensive laws that include a GHG-reduction target, legal principles, legal institutions, supervisory mechanisms, and legal liability, whereas other countries – such as Japan, South Korea, and the Philippines – do not include a reduction target within their climate change laws.[11]

The framework legislative model has gained wide support among Chinese scholars, mainly due to its strength in coordinating various interests.[12] The draft proposal produced by CASS is more of a framework than a comprehensive law. Some scholars maintain that this model is an improvement on the existing legal framework on climate change. For example, Pan argues that the current laws are piecemeal, ambiguous, and not proactive; therefore, a framework law on climate change is essential to clarify principles, institutions, and mechanisms for responding to it.[13] Wang Mingyuan and Wang Shekun acknowledge that China's laws, such as the Renewable Energy Law and LPCAP, have contributed to GHG reduction, yet this non-integrated legislative approach is unable to respond to the needs of addressing climate change.[14] Wang and Liu emphasize that given China's economic and technological constraints, setting binding reduction targets in the law will put too much burden on industry and will lead to implementation difficulties; hence, they recommend a framework legislation to establish national climate change policy and incentive mechanisms.[15] There has been much literature discussing the climate change legislation and experience of other countries, most of

8 Chen Haisong, 'Responding to Climate Change in Developing Countries: Legislation and Revelation' (发展中国家应对气候变化立法及其启示) (2013) 4 *Journal of Nanjing University of Technology (Social Science edition)* (南京工业大学学报(社会科学版)) 5–14.

9 Ibid.

10 Wang Canfa and Liu Zhe, 'The Legislation Model of Responding to Climate Change in China' (论我国应对气候变化立法模式的选择) (2015) 6 *Journal of China University of Political Science and Law* (中国政法大学学报) 113–121.

11 Ibid.

12 Gong Wei, 'International Law on Climate Change and the Legislative Model of China's Climate Change Response' (气候变化国际法与我国气候变化立法模式) (2013) 37(3) *Journal of Xiangtan University* (湘潭大学学报) 42–46.

13 Pan Xiaobin, 'The Construction of China's Legal System on Climate Change' (中国应对气候变化法律体系的构建) (2016) 6 *Nankai Journal (Philosophy and Social Science Edition)* (南开学报(哲学社会科学版)) 78–85.

14 Wang Mingyuan and Wang Shekun, 'Making A Specialized Climate Change Law' (制定专门的气候变化基本法) *Chinese Social Science Newspaper* (中国社会科学报) (8 December 2011).

15 Wang and Liu (n 10).

which maintains that what is most suitable for the Chinese context is a climate law that establishes institutions and principles, defines government responsibilities, and clarifies key mechanisms, such as information disclosure.[16]

Li Yanfang and colleagues argue that the optimal choice for China is to formulate a comprehensive climate law that covers the target, principles, and institutional setting of addressing climate change, legal mechanisms of mitigation and adaptation, and international cooperation on climate change. However, Li et al. also acknowledge that formulating such a comprehensive law would encounter great difficulties because it engages with large areas, sectors, and interests. Insufficient theoretical and empirical research on adaptation also increases the difficulty of making such legislation. They propose that the second-best choice would be to develop a low-carbon development promotion law and a climate change adaptation law to address mitigation and adaptation, respectively. The third-best choice would be to make lower-level regulations, such as on carbon-emission trading on the basis of China's pilot ETS (emission-trading scheme) programmes.[17] Yet in another article by Li Yanfang, she makes a different proposal that specific laws through LPCAP are feasible for China to achieve synergetic effects in air-pollution control and carbon-emission reduction.[18] It shows that some scholarly research lacks consistency.

Instead of choosing one model or the other, some scholars try to sketch a picture of the ideal legislative framework. Zhang Zitai observes that China is responding to the international trend of shifting from climate policy to climate legislation.[19] He proposes that climate legislation should be developed by setting up a framework law and then formulating a mitigation law and adaptation law. The framework law would prescribe key legal principles and institutions for addressing climate change; the mitigation law would cover energy savings and emission reductions,

16 For example, Luo Li, 'Climate Change Legislation in Japan' (日本应对气候变化立法研究) (2010) 5 *Law Review* (法学论坛) 107–113; Song Xixiang and Gao Dali, 'Analysis of UK's Climate Change Act and Its Reference to China' (论英国《气候变化法》及其对我国的启示) (2011) 2 *Journal of Shanghai University (Social Science Edition)* (上海大学学报(社会科学版)) 87–98; Shao Daoping, 'Foreign Legislation on Climate Change and Its Inspiration for China' (国外应对气候变化立法及对中国的启示) (2010) 17 *Environmental Protection* (环境保护) 66–68; Li Yanfang, 'Comparison on National Legislation on Climate Change and Its Implications for China' (各国应对气候变化立法比较及其对中国的启示) (2010) 4 *Journal of Renmin University of China* (中国人民大学学报) 58–66; Yu Wenxuan and Tian Danyu, 'Climate Change Legislation in the US and Mexico and Its Implications for China' (美国和墨西哥应对气候变化立法及其借鉴意义) (2016) 18(2) *Journal of Jiangsu University (Social Science Edition))* (江苏大学学报(社会科学版)) 1–6.

17 Li Yanfang et al., 'Thoughts on China's Climate Change Legislation' (我国应对气候变化立法的若干思考) (2016) 1 *Journal of Shanghai University (Social Science Edition)* (上海大学学报(社会科学版)) 1–12.

18 Li Yanfang and Tian Shiyu, 'Climate Change Regulatory Risk in the Context of Uncertainty and Complexity' (不确定性与复杂性背景下气候变化风险规制立法) (2018) 58(2) *Journal of Jilin University for Social Science* (吉林大学社会科学学报) 42–50.

19 Zhang Zitai, 'Preliminary Study of China's Legal System on Climate Change' (中国气候变化应对法框架体系初探) (2010) 5 *Journal of Nanjing University* (南京大学学报) 37–43.

industrial restructuring, and the circular and low-carbon economy; and the adaptation law would cover natural resources conservation, climatic-disaster reduction, and social security.[20]

Li Yumei supports the idea of developing a comprehensive climate change law at the national level while amending or updating existing laws related to climate change to ensure that they are consistent and contribute to mitigation and adaptation.[21] A crucial point, in Li's argument, is to combine command-and-control and market mechanisms, in other words coercive and incentive measures. Pan goes further in terms of developing supporting regulations. He suggests establishing a three-tiered legislation. The first tier would include a basic climate change law clarifying legislative purpose, legal principles, institutional settings, and strategic planning. The second tier would cover mitigation and adaptation law. The third tier would be concerned mainly with regulations on technology, finance, and capacity building to support both mitigation and adaptation. According to Pan's proposal, the crucial point in developing a climate change legal framework is the legislation on emission trading.[22]

Wang Mingyuan and Wang Shekun, although not laying out a legislative structure as other scholars have done, also propose that crucial issues of climate change law should be clarified in legislation, such as China's climate change strategy, the role of government, a carbon tax, and emission trading. They argue that the implementation of these elements needs to be sustained by specialized regulations and rules about collecting statistics, reporting, reducing carbon intensity, and building adaptive capacity.[23] Most scholarly works on the legislative model are rhetoric more than reality, as they fall short on empirical research on why one model works better than another.

There is considerable agreement among scholars on which principles should be embedded in climate change legislation. One of the representative articles suggests that climate law should include the principles of common but differentiated responsibility (CBDR), public participation, transparent information, and international cooperation.[24] Lü and Liang argue that protecting national interests should be the guiding principle for China's participation in the international climate change governance regime and for the development of its domestic climate change law regime.[25] On their part, both Wang Canfa and Cao Mingde emphasize the importance of climate justice, a key concept for distributing the interests and

20 Ibid.

21 Li Yumei, 'Study of China's Climate Change Legislation' (我国气候变化立法刍议) (2015) 33(1) *Tribune of Political Science and Law* (政法论坛) 182–187.

22 Pan (n 13).

23 Wang and Wang (n 14).

24 Zhao Jun, 'Legal Principles of Climate Change Legislation' (我国应对气候变化立法的基本原则研究) (2015) 7 *Politics and Law* (政治与法律) 80–86.

25 Lü Jiang and Liang Xiaofei, 'Core National Interests and Climate Change Legislation: between Principles and Norms' (国家核心利益与气候变化立法：在原则与规范之间) (2014) 35(4) *Journal of Jishou University (Social Science Edition)* (吉首大学学报(社会科学版)) 34–40.

burdens of climate change, in the design of climate law. Climate justice entails giving adequate attention to the interests of the most vulnerable.[26] According to Wang and Cao, when drafting such a law, climate justice should be embodied in the legislative purpose and legal principles. Cao also maintains that climate justice should be China's tenet when participating in the international climate change regime and responding to climatic challenges at the domestic level. To achieve climate justice, China should take legal, market, and policy measures to reduce GHG emissions and advance the process of making a specialized climate law.[27] Scholars in this area seem to use "principle" in a broad sense, to include both legal principles and policy or guiding principles. For example, many scholars propose "attaching equal importance to mitigation and adaptation" as a legal principle in climate law. There is insufficient discussion on how different legal principles would inform the content of a climate law in terms of mechanisms, liabilities, and so on, for climate change mitigation and adaptation.

In terms of the level at which legislation should be passed, Wang Hui, writing in 2017, explains that China did not make a national-level climate law prior to the Paris Agreement because the country was not legally obligated to reduce its level of GHG emissions under the UNFCCC.[28] As a global and national issue, climate change entails legislation at an international and national level, which suggests that provincial-level legislation will be less effective than national legislation will be.[29] However, Wang concedes that climate change legislation at the provincial level could work as a test case for China's climate change law at the national level, and therefore, it is important that the central government create incentives for local governments to make local-level climate change regulations.[30] Pan argues that the difficulty of enacting a national-level climate change law is largely due to insufficient legislative experience in the area of climate change and divergent interests among the country's 31 provinces.[31] He suggests that local-level legislation should be pioneering in providing experiences for central-level legislators with which to make climate law. Furthermore, local-level legislation is a key step in transforming China's Nationally Determined Contribution (NDC) under the Paris Agreement to locally determined contributions, responding to impacts experienced at local level, and achieving synergetic effects between air-pollution

26 Wang Canfa and Chen Yijian, '"Climate Justice" and China's Climate Change Legislation: Legislative Goal and Institutional Choice' ("气候正义"与中国气候变化立法的目标和制度选择) (2014) 2 *Social Science and Chinese Universities* (中国高校社会科学) 125–139.

27 Cao Mingde, 'The Legal Position and Strategy for China's Participation in Global Climate Governance' (中国参与国际气候治理的法律立场和策略-以气候正义为视角) (2016) 1 *China Legal Science* (中国法学) 29–48.

28 Wang Hui, 'Local Level Climate Legislation in the US and Its Implications for China' (美国地方气候变化立法及其启示) (2017) 17(1) *Journal of China University of Geosciences (Social Science Edition)* (中国地质大学学报(社会科学版)) 56–64.

29 Ibid.

30 Ibid.

31 Pan Xiaobin, 'Study of Local Level Legislation on Climate Change in China' (中国地方应对气候变化先行立法研究) (2017) 3 *Law Science Magazine* (法学杂志) 132–140.

control and carbon-emission reduction.[32] At the same time, he emphasizes that local-level legislation entails the enlargement of local legislative authority, the prevention of carbon leakage and local protectionism, and building legislative coordination among provinces sharing the same concern.[33]

Besides the aforementioned normative scholarship on which level of legislation is more effective and feasible for the response to climate change, some scholars have carried out empirical studies on the efficacy of current experimental provincial climate change regulations.[34] Peng and He find that provincial climate change legislation is generally more policy oriented, lacking regulatory rules and legal liabilities that could be implemented. Li Yumei analyses the institutional difficulties of local-level climate change legislation and shows that it faces challenges in terms of inadequate national legal support and empowerment, ambiguous power distribution, and insufficient financial or technological support.[35]

Can CO_2 be regulated as an air pollutant and through air-pollution control law?

The question of how to reduce GHGs has been given high priority in the Chinese scholarly debate. There are generally three approaches to climate change mitigation: emission trading, carbon taxation, and direct governmental regulation through setting, for example, energy-saving targets. The literature deals mainly with two questions on GHG reduction: Could CO_2 be regulated as an air pollutant? And is carbon trading or a carbon tax the better choice for China?

A few scholars propose to define CO_2 as an air pollutant. Tang uses the US experience to support her argument that identifying CO_2 as an air pollutant could facilitate the application of an environmental impact assessment (EIA) to projects emitting GHGs, and it provides a legal basis for GHG regulation.[36] Yao argues that the absence of a specialized law on GHG emissions allows existing law to fill in, and LPCAP is a good option due to its total-quantity control mechanism.[37] Meng suggests integrating GHG reduction into LPCAP under a separate chapter of the law and giving environmental authorities the role of setting GHG-emission standards.[38]

32 Ibid.

33 Ibid.

34 Peng Benli and He Yan, 'Climate Change Legislation in China: Features, Limitations and Prospects' (中国应对气候变化地方立法 – 特点、局限与展望) (2017) 6 *Journal of Yulin Normal College (Philosophy and Social Science Edition)* (玉林师范学院学报(哲学社会科学)) 65–69.

35 Li (n 21).

36 Tang Shuang'e, 'GHG as An Air Pollutant: Debate in the US and Its Implications for China' (美国关于温室气体为"空气污染物"的争论及对我国的启示) (2011) 4 *Journal of Environmental Management College of China* (中国环境管理干部学院学报) 1–4.

37 Yao Ying, 'The Approach of Regulating GHGs through the Law on Prevention and Control of Atmospheric Pollution' (《大气污染防治法》规制温室气体排放的进路探析) (2014) 18 *Environmental Protection* (环境保护) 55–57.

38 Meng (n 4).

Most scholars, however, object to regulating CO_2 as an air pollutant or using the LPCAP to do so. Li Yanfang and Zhang Zhongli emphasize the fundamental distinction between air-pollution control and CO_2 reduction.[39] Traditional command-and-control measures produce limited effects for CO_2 reduction because the latter is normally addressed by energy-efficiency improvements and low-carbon development through the adjustment of industrial and energy structures.[40] Therefore, China could not simply copy the US's experience of defining CO_2 as an air pollutant.[41] Chang Jiwen argues that if CO_2 were recognized as an air pollutant under Chinese law, developed countries would put pressure on China to use environmental law to set up a CO_2 emission standard.[42] His point is that this would harm China's industrial development.[43] Hu and Zheng concur that including CO_2 in LPCAP could place China in a disadvantageous position at the international level.[44] Zhao and Sun deem the regulation of CO_2 by LPCAP undesirable mainly because of the potential implementation difficulties. They suggest that regulating CO_2 through market mechanisms and administrative incentives is the more practical solution.[45]

The LPCAP, as amended in 2015, requires the coordinated control of air pollution and GHG emissions. However, the legal status of CO_2 remains unclear, leaving much space for debate for scholars. The 2019 institutional reform transferring responsibility for dealing with climate change from the NDRC to the MEE has not been fully discussed by scholars in terms its effects on climate change governance. While the legal status of CO_2 is still uncertain, the question of how to produce synergetic effects between the control of air pollution and CO_2 is a challenging topic for China.[46]

39 Li Yanfang and Zhang Zhongli, 'Legal Status of CO_2 and the Legislative Approach to Its Emission Regulation' (二氧化碳的法律定位及其排放规制立法路径选择) (2015) 2 *Social Science Study* (社会科学研究)108–113.

40 Ibid.

41 Ibid.

42 Chang Jiwen, 'The Emission Control of CO_2 and the Amendment of the Law on Prevention and Control of Atmospheric Pollution' (二氧化碳的排放控制与《大气污染防治法》的修订) (2009) 3 *Law Science Magazine* (法学杂志) 74–76.

43 Ibid.

44 Hu Yuan and Zheng Shaohua, 'From Authoritarian Management to Social Governance: Reflections on the Amendment of the Law on Prevention and Control of Atmospheric Pollution' (从威权管制到社会治理 – 关于修订《大气污染防治法》的几点思考) (2010) 6 *Modern Law Science* (现代法学) 150–156.

45 Zhao Jun and Sun Songlong, 'An Assessment of Integrating GHGs within the Law on Prevention and Control of Atmospheric Pollution' (《大气污染防治法》将温室气体排放纳入控制体系的评析) (2016) 33(1) *Journal of Shanghai University (Social Science Edition)* (上海大学学报(社会科学版)) 27–34.

46 Gong Wei, 'Cooperative Control of Air Pollutants and GHGs: Challenges and Responses from the Perspective of Enforcement' (大气污染物与温室气体协同控制面临的挑战与应对 – 以法律实施为视角) (2017) 1 *Journal of Southwestern University for Nationalities (Humanities and Social Science Edition)* (西南民族大学学报(人文社会科学版)) 108–113.

Which regulatory instrument works better for China in mitigating GHGs?

The Chinese government initiated a pilot ETS programme in 2012, before the costs and benefits of the carbon market or carbon tax were fully understood in China. Pilot projects of carbon trading in selected provinces were to be the key strategy to test this choice. However, the Chinese government's stance on the carbon tax remains ambivalent. Scholars note the possibility that the government will impose a carbon tax on emitters of below the equivalent of 5,000 tons of standard coal emissions (the minimum for ETS liability) or on those that are outside the scope of ETS coverage.[47] Yet there has been no movement towards the adoption of a carbon tax in China. The country's finance minister has asserted that China would not set a separate carbon tax but would integrate it into the existing environmental tax or resource tax.[48] The debate between a carbon market and carbon tax still persists among scholars, despite the fact the central government has expressed its preference for a carbon market.

Liu Mingming argues that when governments regulate GHG emissions, they should conduct a cost-benefit analysis, distinguish survival emissions from luxury emissions, and consider the government's regulatory capacity and efficiency.[49] Xie finds that although many economists prefer a carbon tax as a GHG-reduction tool due to its high efficiency compared to carbon trading, developed countries or economic entities such as the European Union choose carbon trading due to its advantages in overcoming political barriers, improving global reduction efficiency, and providing incentives for developing countries to cooperate.[50] Xie, therefore, urges that when making its own choice, China should give full consideration to what developed countries have done.

Cao Mingde makes a case for using emission trading for large emitters and sectors while imposing a carbon tax on small and medium emitters.[51] Zheng and Dou argue that compared to a carbon tax, carbon trading comes with many fewer negative effects on the economy and industry, with much-more-certain reduction effects, wider acceptance among emitters, and broader international influence.[52]

47 Zhu Yan, 'Asking again about Carbon Reduction, Will There Be Carbon Tax?' (再问碳减排，何日"税"会来) *China Energy Newspaper* (中国能源报), 22 October 2018.

48 Yan Lu and Du Yanfei, 'China Will Not Establish A Separate Carbon Tax' (People Net, 21 March 2016) <http://energy.people.com.cn/n1/2016/0321/c71661-28214363.html> (accessed 29 October 2019).

49 Liu Mingming, 'The Regulation of GHG Emissions: Content and Boundary' (论温室气体排放管制的内涵及边界) (2012) 18 *Journal of Central South University (Social Science Edition)* (中南大学学报(社会科学版)) 105–111.

50 Xie Laihui, 'Carbon Emission Trading or Carbon Tax: Theory and Policy' (碳交易还是碳税? 理论与政策) (2011) 6 *Financial Review* (金融评论) 103–110.

51 Cao (n 27).

52 Zheng Shuang and Dou Yong, 'Coping with Climate Change with Economic Instruments: Comparison Between Carbon Tax and Carbon Trading' (利用经济手段应对气候变化 – 碳税与碳交易对比分析) (2013) 35(10) *China Energy* (中国能源) 11–15.

Furthermore, they find that the social and economic cost and uncertainty of creating a carbon tax is high, and therefore, a carbon-trading approach is a better choice for China.[53] For scholars advocating carbon trading, a critical issue that has not been fully discussed is how this mechanism can operate well in a developing socialist market characterized by manifold government interventions in the industry and energy sectors.

Hu and Gao argue that China's market economy is immature, and therefore the establishment and operation of carbon trading will face risks.[54] A carbon tax, by contrast, brings little risk to the market and is more compatible with China's current legal system. They also argue that enforcing a carbon tax will be cheaper than establishing a carbon market in China.[55] Wang Mingyuan adds that a carbon tax is more feasible for China compared to a carbon market because the latter entails much-more-refined administrative management tools, such as total-quantity control, quota distribution, monitoring and verification, and regulating allowance trading.[56] Deng finds that implementing a carbon tax in China may face several challenges, including incompatibilities with existing tax regimes, the ETS system, and international trade laws. He therefore suggests, in the rather inconclusive style that characterizes some scholarship in China, that the differences between a carbon tax and pollution-discharge fees and between a carbon tax and carbon trading should be studied thoroughly before adopting and implementing a carbon tax in China.[57]

Some scholars focus on developing a coherent system of carbon regulation with diverse instruments, identifying coordination between a carbon market and a carbon tax as an important issue. Zhang Zhongli suggests that emission trading is limited in achieving the aim of adjusting energy and industrial structure and changing public behaviour. It therefore must be supplemented by other methods. One such method is the regulatory approach towards emission sources that are not covered by the emission-trading scheme, such as the regulation of renewable-energy development (including a renewable-energy quota system), carbon tax, and performance standards on carbon emissions.[58] The second method proposed by Zhang is an incentive approach to low-carbon development, such as a low-carbon-product standard, authentication, and labelling policy and a low-carbon supply-chain management. According to Zhang, it is important to coordinate the

53 Ibid.

54 Hu Bin and Gao Fang, 'Carbon Trading or Carbon Tax?' (碳交易还是碳税) (2015) 4 *World Environment* (世界环境) 50–52.

55 Ibid.

56 Wang Mingyuan, 'Climate Change: An Environmental Problem or a Developmental Problem' (气候变化,是环境问题还是发展问题?) (2016) Z1 *Environmental Economics* (环境经济) 42–43.

57 Deng Haifeng, 'Legal Mechanisms for Implementing the Carbon Tax' (碳税实施的法律保障机制研究) (2014) 4 *Global Law Review* (环球法律评论) 104–117.

58 Zhang Zhongli, 'Legal System Construction of GHG Emission Control' (论温室气体排放控制法律制度体系的建构) (2016) 2 *Tsinghua Forum of Rule of Law* (清华法制论衡) 382–401.

ETS with energy-related regulations, a carbon tax, and the air-pollution prevention law so as to improve synergetic effects.[59]

Key legal issues in carbon trading

An ETS is a complex system that involves broad and diverse legal issues, including the legal nature of carbon-emission rights, approaches to setting up the emission cap and the allocation of allowances under the cap, and rules on monitoring, reporting, and verification (MRV).[60]

Some scholars have tracked the regulatory development of carbon trading and assessed its effectiveness in China. They all agree that the current regulatory framework is insufficient to provide solid support for carbon-market activities and that a higher-level regulation or law with detailed rules is the solution. He Ying concludes that although there have been some rules on carbon trading both at national and local level, they are normative documents with a low legal hierarchy and limited binding effect.[61] Therefore, He suggests that a higher-level regulation or law on carbon trading is desirable. Wang Binhui finds that the legal system on carbon trading is still in its infancy due to the undefined nature of emission allowances, unfair allowance allocation, and problems in relation to the MRV system.[62] Wang proposes to make a specialized carbon-trading law to define the legal nature of carbon-emission rights, clarify the reduction obligations of emitters, establish supervision mechanisms for carbon trading, and clarify emitters' liabilities.[63] Shi and Yang analyse the penalties for liable entities that fail to comply with the requirements under the ETS pilots, finding that current penalties do not encourage emitters to comply. They therefore suggest monetary and credit-record penalties in the provisions of the national regulation on carbon trading.[64] Liu Siqi assesses the implementation of ETS pilots, arguing that the current carbon market suffers from limited coverage of emitting enterprises and GHGs, overreaching administrative intervention, inconsistent regulations, and limited attention to the

59 Ibid.

60 Chen Huizhen, 'Reduction Target and Total Emission Allowance: Experience and Implications of the EU's Carbon Emission Scheme' (减排目标与总量设定：欧盟碳排放交易体系的经验及启示) (2013) 15(4) *Journal of Jiangsu University (Social Science Edition)* (江苏大学学报(社会科学版)) 14–23.

61 He Ying, 'Legislative Regulation of Carbon Emission Trading' (我国碳排放权交易立法规制思考) (2018) 2 *Journal of Southern China Normal University (Social Science Edition)* (华南师范大学学报(社会科学版)) 157–161.

62 Wang Binhui, 'The Development of Carbon Emission Trading and Its Legislative Follow-up' (我国碳排放权交易的发展及其立法跟进) (2015) 13(2) *Present-day Law Science* (时代法学) 13–25.

63 Ibid.

64 Shi Xueying and Yang Bowen, 'Penalty Rules of Regulated Enterprises Failing to Achieve Performance Targets' (控排企业碳交易未达履约目标的罚则设定) (2018) 28(4) *China's Population, Resources and Environment* (中国人口·资源与环境) 35–42.

achievement of the reduction goal.[65] A higher-level law to regulate the carbon market, he suggests, should clarify the responsibility for supervision and define the role of administrative management in the carbon market.

In terms of the legal status and nature of emission allowances, there are diverse opinions among scholars and the disagreement generally centres on whether the nature of allowances is more government oriented or more market oriented. The definition influences how the legal system on carbon trading should be designed. Both Deng Haifeng and Wang Mingyuan regard emission rights as a quasi-property right, largely due to the scarcity and certainty of atmospheric capacity under the UNFCCC and Kyoto Protocol and due to their disposability and tradability.[66] Ding and Pan agree to define a carbon-emission allowance as a quasi-property right, but they also identify it as an environmental right, representing the economic value and the ecological value, respectively.[67] Although defining a carbon-emission right as a property right has great benefits in optimizing the allocation of climate resources, there is concern about the risk of transformation of public resources into freely gained private assets under the current allowance-allocation system.[68] Zhang Lei notes that the atmospheric capacity is constantly changing and hard to quantify, which makes the rule of a property-right system hard to accept.[69] Yang Zewei takes the view that the carbon-emission right is a new right for development, with its legal foundation in the UNFCCC, the Kyoto Protocol, the principle of sustainable development, CBDR, and the principles of fairness and justice.[70] The distribution of carbon emissions as a development right, therefore, should consider the development need, population, historical responsibility, fairness and equality, and the level of energy efficiency.[71] Wang Hui argues that a carbon-emission right is better defined as a concession right, which is granted by administrative agencies to emitters.[72] According to his argument, this definition enables the

65 Liu Siqi, 'The Status Quo, Problem Analysis and Countermeasures Study of China's Carbon Emission Trading' (中国碳排放交易试点的现状、问题分析及对策研究) (2015) *China's Environmental Rule of Law* (中国环境法治·2015年卷) 27–38.

66 Wang Mingyuan, 'Carbon Emission Right as a Quasi-Property Right and Developmental Right' (论碳排放权的准物权和发展权属性) (2010) 6 *China Legal Science* (中国法学) 92–99; Deng Haifeng, 'Systematic Outline of the Pollution Emission Right' (排污权制度论纲) (2007) 6 *Legal Science* (法律科学) 76–83.

67 Ding Ding and Pan Fangfang, 'Legal Attributes of the Carbon Emission Right' (论碳排放权的法律属性) (2012) 9 *Legal Science Magazine* (法学杂志) 105–108.

68 Zhang Lei, 'GHG Emission Right as Property Right and Its Institutional Dilemma: Refection on Hardin's "The Tragedy of the Commons"' (温室气体排放权的财产权属性和制度化困境 – 对哈丁"公地悲剧"理论的反思) (2011) 1 *Law and Social Development* (法制与社会发展) 101–110.

69 Ibid.

70 Yang Zewei, 'Carbon Emission Right: A New Developmental Right' (碳排放权：一种新的发展权) (2011) 41(3) *Journal of Zhejiang University (Humanities and Social Science)* (浙江大学学报(人文社会科学版)) 40–49.

71 Ibid.

72 Wang Hui, 'Carbon Emission Right as Concession' (论碳排放权的特许权本质) (2017) 6 *Law and Social Development* (法制与社会发展) 171–188.

government to flexibly respond to the uncertainty of climate change by creating and revoking emission rights and avoids administrative compensation.[73]

Emission allowance allocation is a key element of the legal framework related to an ETS. With the purpose of promoting fair competition and corporate compliance, a crucial question for China is how the central and provincial governments will share the authority for allowance allocation. Cheng and Chen suggest that China could combine the decentralized and centralized experience learned from the EU ETS, where the regional government sets allocation principles, rules, and methodologies through regulations, and state governments develop their own implementation rules based on their local circumstances.[74] In terms of how to choose among the three allocation methods, known as free allocation, auction, and mixed allocation (part free, part auction), Xu and Feng conclude that the preliminary stage of carbon trading should be dominated by free allocation while allowance auctions may be more safely applied when the scheme is mature.[75]

Zhu Deli argues that risks produced in the process of emission trading have received little attention among Chinese scholars.[76] She finds that the key risks are unstable transaction prices, poor environmental performance due to an inappropriate allocation methods, weak incentives for technology innovation, and an inadequate support system.[77] Other scholars discuss more-specific issues, such as the MRV systems used by China's pilot programmes. Pang and Yan find that although local governments, namely Beijing, Shanghai, Guangdong, Tianjin, and Shenzhen have published diverse regulations or rules on MRV, they are generally underdeveloped.[78] They suggest that despite the autonomy that local governments have to determine MRV rules, the monitoring and accreditation of verification agencies need to be centralized, and there is also a need to cross provincial boundaries to allow competition between verification agencies so as to develop the market in verification services.[79]

Many scholars have studied the experience of the ETS in other jurisdictions in order to find useful references for China. Peng and Shao argue that the EU ETS is currently the most effective model, and that China should consider linking its

73 Ibid.

74 Cheng Xinhe and Chen Huizhen, 'Two Dimensions of Carbon Trading Allowance: A Case Study of Guangdong Province' (论碳交易配额分配法律制度的两大维度 – 以广东为例) (2014) 2 *Journal of South China Normal University (Social Science Edition)* (华南师范大学学报(社会科学版)) 103–108.

75 Xu Yixiang and Feng Cuihua, 'Initial Allocation of the Carbon Emission Right: Legal Institutions and Content' (碳排放权初始分配法律制度的基本构造) (2013) 7 *Tribune of Study* (学习论坛) 70–73.

76 Zhu Deli, 'The Risk and Legal Response to the Carbon Trading Scheme' (碳排放权交易机制的风险挑战及其法律应对) (2018) 4 *Studies in Dialectics of Nature* (自然辩证法研究) 78–83.

77 Ibid.

78 Peng Feng and Yan Lidong, 'Comparative Study of Monitoring, Reporting and Verification of Local Carbon Trading Pilot' (地方碳交易试点之"可测量、可报告、可核实制度"比较研究) (2015) 15(4) *Journal of China University of Geosciences* (中国地质大学学报) 26–35.

79 Ibid.

ETS with the EU ETS in the future.[80] They suggest that key steps in establishing China's national ETS include clarifying the roles of regulatory agencies, setting up a cap, allocating the allowances, and scoping the coverage of the sectors.[81] However, Li Zhiping warns that the practice of the EU ETS and the carbon market at the state level in the United States show that there exists market abuse, price manipulation, and fraud, even though they have established complex supervision institutions, strict regulations, and regular market investigation and assessment.[82] For China, a comprehensive supervision system should be established, which includes a multilevel supervision system by the governments, third-party agencies, carbon-emission exchanges, and society.[83] At the same time, she proposes to promote information transparency on carbon-emission trading, prevent market abuse, and clarify liabilities of noncompliance.[84] After examining South Korea's experience in carbon trading, Zhang Zhongli argues for the importance of national planning to create stability and certainty for low-carbon investors.[85] What is also important, he argues, is to clarify the scope of coverage to determine the entities that are liable under the emission cap and their reporting obligations, as well as the trading rules on the secondary market and corresponding liabilities.[86]

The role of the judiciary in addressing climate change has long been ignored in China. The situation changed slightly after the Supreme People's Court in 2016 issued a policy on providing a judicial service for environmental protection.[87] The policy states that the judicial response to climate change is important and that courts should deal with cases in relation to carbon trading. Although there has been no case of this kind, there is some emerging scholarship on the possibility of running administrative, criminal, and civil cases relating to carbon trading.[88] Lü and Wang note that China's ETS pilots were established by local regulations and policies

80 Peng Feng and Shao Shiyang, 'Carbon Emission Trading in EU: New Developments and Implications for China' (欧盟碳排放交易制度：最新动向及对中国之镜鉴) (2012) 12(5) *Journal of China University of Geosciences* (中国地质大学学报) 41–47.

81 Ibid.

82 Li Zhiping, 'Supervision Mechanism of Carbon Trading Market' (碳交易市场的监管机制研究) (2012) 14(1) *Journal of Jiangsu University (Social Science Edition)* (江苏大学学报(社会科学版)) 56–62.

83 Ibid.

84 Ibid.

85 Zhang Zhongli, 'Carbon Trading Law in South Korea and Its Application to China' (韩国碳排放交易法律及其对我国的启示) (2016) 5 *Northeast Asia Forum* (东北亚论坛) 50–62.

86 Ibid.

87 Supreme People's Court, Opinions of the Supreme People's Court on Giving Full Play to Judicial Functions to Provide Judicial Services and Guarantee for Promoting Ecological Civilization Construction and Green Development (最高人民法院关于充分发挥审判职能作用为推进生态文明建设与绿色发展提供司法服务和保障的意见) Supreme People's Court Order No. 12 (2016), 2 June 2016.

88 Pan Xiaobin, 'Litigation of China's Carbon Emission Trading' (我国碳排放交易法律诉讼问题) (2016) 4 *Tianjin Legal Science* (天津法学) 32–39.

before any national law that clarified key legal issues.[89] As a result, judicial dilemmas are pervasive, such as the unclarified legal status of CO_2 (whether it is legally an air pollutant), the undefined legal nature of emission rights, the inadequate judicial remedy for allowance allocation (liable entities are not entitled to sue), insufficient access to the judiciary by regulators and liable entities to resolve disputes (in the event that liable entities do not surrender sufficient allowances), and lack of public oversight (on setting the total emission cap, allowance allocation, and penalty provisions).[90] Lü and Wang suggest regulating CO_2 as an air pollutant and a carbon-emission right as an intangible carbon asset and suggest developing a dispute-resolution mechanism through the judiciary on allowance allocation and trading.[91]

Conclusion

Legal scholarship on climate change is underdeveloped in China when compared with its international counterparts as well as with English-language scholarship on Chinese environmental law by Chinese authors. Although some purely scholarly discussions on legal issues of climate change exist (e.g. the legal status of emission rights), most work attempts (in typical Chinese scholarly fashion) to provide practical advice to the Chinese government. Without frequently referring to secondary sources (i.e. law journal articles and books) in the English language, Chinese legal scholarship on climate change does not adequately engage with the international scholarship on climate law. The reasons are hard to identify, but language is probably a significant factor. The currency of the Chinese scholarship is generally up to date with China's international practices and domestic legal developments.

Yet China's authoritarian regime to some extent affects how research is conducted. It is normally conducted by legal scholars to support the decisions by authorities rather than to question, challenge, or shape the decision-making process. Once a top-down choice (e.g. in favour of an ETS) has been made, the scholarly focus generally concentrates on legal issues arising from this decision. The merit of this system is to identify and solve related legal issues in an efficient way.

In terms of whether a stand-alone climate law is necessary, there is a general consensus among Chinese scholars, though for different reasons. Legal analysis on how climate change has fundamentally challenged China's established legal framework is comparatively weak. Arguments on whether GHGs should be regulated by the legal framework on air pollution are divided, yet most object to the

89 Lü Zhongmei and Wang Guofei, 'The Development of China's Carbon Trading Market: Judicial Issues and Responses' (中国碳排放市场建设：司法问题及对策) (2016) 5 *Gansu Social Science* (甘肃社会科学) 161–168.

90 Ibid.

91 Ibid.

idea of integrating the two because of their fundamental differences. Given the stipulation on the coordinated regulation of GHGs and air pollutants, more analysis is needed on how this could be implemented.

Debates on carbon trading versus a carbon tax are extensive. The question about which regulatory instrument is more compatible with China's existing legal framework has been discussed among legal scholars, but at least the early literature does not definitively support the implementation of an ETS. With China's increasing importance under the UNFCCC processes, there is a strong need for climate law scholars working in China to conduct more research on the legal challenges related to the compliance regime under international climate law. Furthermore, on a positive note, as the number of Chinese young scholars with overseas law degree increases, future research conducted by Chinese scholars is more likely to engage with English-language scholarship in the area of climate law.

3 China in global context

Continuity and discontinuity in international and domestic climate change law and policy

Ying Shen

Introduction

The remarkable rise in greenhouse gas (GHG) emissions in China, caused by more than three decades of energy-intensive and coal-fired economic growth, represents the greatest challenge to the global effort to combat climate change in the coming years. While GHG emissions in developed countries are stable or decreasing,[1] China's energy-related carbon dioxide (CO_2) emissions are estimated to have contributed more than 25% to the global total in 2015[2] and are expected to increase until at least 2030.[3] Therefore, efforts to find a solution to climate change are unlikely to succeed without a significant contribution from China. Meanwhile, with the rise of its economy, China has been transformed into a climate superpower, on a par with the European Union (EU) and the United States (US), capable of making or breaking a global climate deal. The role of China in the context of global climate governance is therefore crucial.

The Chinese government has begun to implement a series of policies and regulations aimed at improving the energy- and carbon-intensity of its economy. For quite some time, China's regulatory approach to the control of GHG emissions has relied heavily on top-down, command-and-control regulations that regulate behaviour through directives on emission-control levels or methods. The enforced closure of inefficient power plants and factories and similar administrative and political measures are used as part of last-ditch efforts to achieve energy-intensity targets set by the government. For example, to meet energy-intensity targets for 2006–2010, a number of Chinese provinces were forced to shut down large swathes of their industrial capacity, and "blackouts" of industries in certain cities were common at the end of 2010.[4] This resulted in a slight rebound of energy

1 Corinne Le Quéré et al., 'Drivers of Declining CO_2 Emissions in 18 Developed Economies' (2019) 9 *Nature Climate Change* 213, 213–217.

2 Enerdata, Global Energy and CO_2 Database (2015).

3 Climate Action Tracker (CAT), 'China – Overview' (Climate Action Tracker, 2 December 2019) <https://climateactiontracker.org/countries/china/> accessed 25 January 2020.

4 See Peng Pu, 'Transforming the Old Mindset and Adjusting the Structure to Reduce Carbon Intensity' (转观念调结构 降低碳强度) *China Energy News* (中国能源报) (Beijing, 17 February 2014)

intensity in early 2011.[5] In addition to the closure of thousands of small inefficient and polluting coal-fired power plants, iron and steel mills, cement kilns, aluminium plants, and so on, China has also tried to increase its share of non-fossil energy through clean and renewable energy, such as hydroelectric, nuclear, and wind power.[6] Other solutions, such as storing carbon in forests, improving the enforcement of standards, giving monetary awards for energy-saving achievements by companies and public institutions, and tightening efficiency standards for buildings and appliances, have also been tried by the Chinese government.[7]

This top-down approach, though significant, is coupled with a broad encouragement of local experimentation and flexibility in implementation.[8] In recent years, the Chinese government has increasingly called for the greater use of market-based instruments to address climate issues. For example, seven pilot emission-trading schemes (ETS) have been implemented in parts of China since 2013. On the basis of experience gained from these schemes, a nationwide ETS was launched in December 2017.[9]

Before the Chinese government's collective framing of its policies on energy and environmental issues as a climate change programme in 2007, its domestic policies (such as afforestation, energy conservation, fuel switching from coal to nuclear power and hydroelectric facilities, energy-price reform, and investment in energy-efficient technologies) were motivated by non-climate objectives,

---

18; Liu Chang, 'Study on the Sustainability of the Effect of Energy-Saving Policy' (节能政策效果的持续性探究) (2014) 5 *Economic Review* (经济纵横) 25, 28; Stephanie Ohshita, Lynn Price, and Tian Zhiyu, 'The Available Target Allocation Method in Chinese Provinces: Energy Intensity During the Twelfth Five-Year Plan Period' (中国省区可用的目标分配方法：第十二个五年规划的能源强度) (Ernest Orlando Lawrence Berkeley National Laboratory (劳伦斯伯克利国家实验室), March 2011) <https://china.lbl.gov/sites/all/files/12th_fyp_target_allocation_report_cn_final_032111_0.pdf> accessed 25 January 2020; Jonathan Watts, 'China Resorts to Blackouts in Pursuit of Energy Efficiency' *The Guardian* (London, 20 September 2010) <www.theguardian.com/world/2010/sep/19/china-blackoutsenergy-efficiency> accessed 25 January 2020; Jane A. Leggett, 'China's Greenhouse Gas Emissions and Mitigation Policies' (CRS Report for Congress No R41919, Congressional Research Service, 2011) 16; Chris Buckley, 'China Eyes Steel, Cement Sectors for Carbon Credit Trade' *Reuters* (London, 4 August 2011) <www.reuters.com/article/2011/08/04/china-emissions-trading-idUSL3E7J407J20110804> accessed 25 January 2020.

5 Chang (n 4), 25–26; Leggett (n 4), 16.

6 For example, the accumulative power generation through hydroelectric power, nuclear power, and wind power during the Eleventh Five-Year Plan period (2006–2010) exceeded 3 trillion kilowatt hours, substituting 1,500 million tons of raw coal. See Wang Youling and Lei Min, 'NDRC: China Has Basically Achieved Its Energy-Saving Targets During the Eleventh Five-Year Plan Period' (国家发改委：中国基本实现"十一五"规划节能降耗目标) (14 January 2011) <www.chinaenergy-council.org/policynews/2011/1/14/2011114134037.html> accessed 25 January 2020; Leggett (n 4), 17.

7 Leggett (n 4), 17.

8 Alex L. Wang, 'Climate Change Policy and Law in China', in Cinnamon P. Carlarne, Kevin R. Gray, and Richard G. Tarasofsky (eds), *The Oxford Handbook of International Climate Change Law,* 635, 637 (Oxford University Press, 2016).

9 For more information on the ETS in China, see Chapter 5 of this book.

such as economic development, energy security, social stability, and international reputation.[10] During China's early engagement in global climate governance, from 1989 to 2007, climate change issues had low domestic priority. Since 2007, the situation has changed, as the Chinese government began to frame its actions as being part of a comprehensive National Climate Change Program.[11] This programme has attempted to mitigate GHG emissions through reduced reliance on energy-intensive heavy industry; greater energy efficiency and conservation; increased use of non-fossil energy and fossil fuels of lower carbon intensity; and the expansion of carbon sinks (e.g. afforestation). This transition has been influenced by China's engagement in global climate negotiations since the United Nations (UN) Climate Change Conference in Bali in 2007. The priority of climate change was further enhanced during the 12th Five-Year Plan period (2011–2015), when the Chinese government shifted to the so-called New Normal model of economic development. Under this model, the government has focused on higher-quality growth (as opposed to energy-intensive growth), which implies that the growth in emissions from the use of coal is likely to slow in the coming years.[12]

This chapter will identify and discuss the continuities and discontinuities between international climate change law and policy on one hand and China's domestic equivalents on the other. It will do so by following a chronological presentation of China's engagement with the international climate change negotiations. The analysis follows Xu and Zhang in dividing China's engagement with the global regime into three phases:

1 A learning phase from 1989 to 1995.
2 A shift towards a more active participation between 1995 and 2007.
3 A more comprehensive engagement in climate change negotiations starting around the time of the Bali Conference in 2007.[13]

This division corresponds with the most important developments in the climate change regime itself. The first period saw China's participation in the early global climate change negotiations leading to the United Nations Framework Convention on Climate Change (UNFCCC). In the second period, China was engaged in the negotiations leading to the 1997 Kyoto Protocol. The third period marks China's comprehensive engagement with the international negotiations during the post-Kyoto (Paris Agreement) era.

10 Wang (n 8), 636.
11 Ibid.
12 Fergus Green and Nicholas Stern, 'China's Changing Economy: Implications for its Carbon Dioxide Emissions' (2017) 17(4) *Climate Policy* 423, 423–442. For more information on the 'New Normal' slogan, see below.
13 Xu Nan and Zhang Chun, 'What the World is Getting Wrong about China and Climate Change' (China Dialogue, 18 February 2013) <www.chinadialogue.net/article/show/single/en/5711-What-the-world-is-getting-wrong-about-China-and-climate-change> accessed 25 January 2020.

Continuity and discontinuity in China's climate change law and policy

Phase I (1989–1995)

China's participation in the UN Conference on the Human Environment (UNCHE) in Stockholm in 1972 marked China's entry into a dialogue on international environmental issues. When a rapid movement towards international negotiations on a climate change convention pushed climate change onto the political agenda, at the end of the 1980s, China again joined the global process.[14]

Before engaging in the early climate change negotiations leading up to the UNFCCC, China had begun to monitor developments in global warming in the scientific community. In 1987, the National Climate Committee was established by the then State Science and Technology Commission (SSTC).[15] Interdisciplinary studies on climate change were also undertaken with the support of the Academy of Sciences and the then National Environmental Protection Agency (NEPA).[16] In the late 1980s, the State Meteorological Administration (SMA) was assigned to coordinate China's involvement with the Intergovernmental Panel on Climate Change (IPCC).[17] In February 1990, the National Climate Change Coordination Group (NCCCG), created under the State Council's Environmental Protection Commission (EPC),[18] was chaired and administrated by the SSTC.[19] The NCCCG had four working groups, with three groups paralleling the organization of the IPCC (scientific assessment led by the SMA, impact assessment led by NEPA, and

14 Ibid.

15 Articles of National Climate Committee (国家气候委员会章程), National Climate Committee (adopted on 6 February 1987), art. 1. The State Science and Technology Commission was reorganized as the Ministry of Science and Technology in 1998. For more information on the history of China's current Ministry of Science and Technology, see www.gjgwy.org (公考资讯网), 'Ministry of Science and Technology' (科学技术部) <www.gjgwy.org/zhongyang/xzjg/11573.html> accessed 25 January 2020.

16 Michael T. Hatch, 'Chinese Politics, Energy Policy, and the International Climate Change Negotiations', in Paul G. Harris (ed), *Global Warming and East Asia: The Domestic and International Politics of Climate Change,* 43, 48 (Routledge, 2003). The National Environmental Protection Agency (a sub-ministerial level agency) was upgraded to the State Environmental Protection Administration (SEPA, Ministerial level) in 1998. The SEPA was upgraded to the Ministry of Environmental Protection (MEP), as an integral department of the State Council, in 2008. The MEP was superseded by the Ministry of Ecology and Environment (MEE) in 2018. For more information on the history of China's current MEE, see MEE, 'History' <http://english.mee.gov.cn/About_MEE/History/> accessed 25 January 2020.

17 China Meteorological Administration, 'Brief Introduction to the State Meteorological Administration' (中国气象局简介) <www.cma.gov.cn/2011zwxx/2011zbmgk/201110/t20111026_117793.html> accessed 25 January 2020.

18 The Environmental Protection Commission (EPC) of the State Council was disbanded in 1998. See MEE (n 16).

19 Hatch (n 16), 49.

response strategies led by the SSTC) and a fourth group focusing on international climate change negotiations (led by the Ministry of Foreign Affairs – MOFA).[20]

Although China's initial interest in the climate change issue was largely scientific, the SSTC and the SMA, whose primary tasks were research on climate change, became increasingly marginalized in China's internal climate change debate when the issue of climate change moved from the scientific to the political stage.[21] In their place, agencies with an economic focus, such as the then State Planning Commission (SPC)[22] and the Ministry of Energy, gradually became involved in global negotiations.[23] In 1990, officials from NEPA, the SPC, the Ministry of Energy, and MOFA met to deliberate on China's international negotiating stance in preparation for the Second World Climate Conference held later that year and for the climate change negotiations due in February 1991.[24] This internal policymaking process consisted of scientific, energy, economic, and political debates among officials with their own priorities and preferences.[25] NEPA accepted the scientific claims about the potential threat of global warming and showed its willingness to take action to limit GHG emissions, whereas the SPC, the Ministry of Energy, and MOFA were more suspicious of the science of global warming, seeing economic development as the highest priority and opposing any type of commitment (such as restructuring the energy sector or limiting GHG emissions) that might impede economic growth.[26] Because the expansion of the country's energy supply would facilitate economic growth, the three agencies wanted to preserve the dominant role that coal played in China's energy mix, thus opposing NEPA's efforts to limit GHG emissions.[27] The three agencies' views were further reinforced by the national leadership's concerns about continued economic

20 Ibid.
21 Ibid., 50.
22 The State Planning Commission was reorganized as the State Development and Planning Commission (SDPC) in 1998. The SDPC was reorganized as the National Development and Reform Commission (NDRC) in 2003. For more information about China's current NDRC, see China.com.cn (中国网), 'National Development and Reform Commission' (国家发展和改革委员会) *China.com.cn* (3 June 2006) <www.china.com.cn/policy/txt/2006-06/03/content_6228593.htm> accessed 25 January 2020.
23 Wang (n 8), 640; Hatch (n 16), 50.
24 Hatch (n 16), 49.
25 The NEPA was in charge of domestic environmental issues. The SPC had its highest priority in maximizing economic development. The primary task of the Ministry of Energy was expanding energy supply to fuel economic growth. The MOFA focused on ensuring China's state sovereignty and independence in pursuing rapid economic development in the context of international negotiations. Information on China's internal dynamics of the political process regarding climate change issues in early 1990s are based on Economy's work; see Elizabeth Economy, 'Chinese Policy-making and Global Climate Change: Two-front Diplomacy and the International Community', in Miranda A. Schreurs and Elizabeth Economy (eds), *The Internationalization of Environmental Protection,* 19–41 (Cambridge University Press, 1997); see also Hatch (n 16), 49–50.
26 Ibid.
27 Ibid.

development.[28] After this internal deliberation, the SPC and MOFA began to dominate China's key negotiation positions in the meetings leading to the UNFCCC.[29]

Although there were different views among officials involved in formulating China's position in the international climate change negotiations, the Chinese delegation presented a unified front to the outside world once the negotiations had started. In the course of 1991, before the conclusion of the UNFCCC negotiating process, China held to the following tenets as its position in the negotiations:[30]

1 Countries should accept common but differentiated responsibilities for responding to climate change.
2 Countries should develop comprehensive and effective cooperation with other countries on the basis of the principle of equity, without infringing on the sovereignty of signatories.
3 Economic development is a necessary precondition for taking measures to tackle climate change. Any measures should give thorough consideration to each country's per capita emissions, ensuring proper levels of consumption for developing countries.
4 Developed countries should on a preferential basis provide developing countries with financing and technology transfer.

From this position, it may be gleaned that state sovereignty and economic development were top priorities for China at the climate change negotiations. The negotiations leading to the UNFCCC provided the Chinese government with an opportunity to raise its international reputation. It acted for the most part as a moderator.[31] For example, when it seemed that the Group of 77 developing countries (G-77) might block the final agreement on a convention, China sought to harmonize different positions within the group and pursued compromise.[32] In the end, positions taken by China (and developing countries in general) were partially acknowledged in the UNFCCC text: developed and developing countries had different responsibilities (however, developing countries were unable to get the principle of main responsibility on the part of developed countries for past GHG

28 During this period, statements from Deng Xiaoping and Li Peng always stressed the importance of rapid economic growth and neglected environmental issues. See Economy (n 25), 29.

29 Hatch (n 16), 50.

30 Fourth Group of National Coordination Committee on Climate Change (国家气候变化协调小组第四工作组), 'China's Preliminary Draft of International Convention on Climate Change' (《关于气候变化的国际公约》条款草案（中国的建议）) in Secretariat of Environmental Protection Commission of the State Council (eds), (国务院环境保护委员会秘书处编) *Corpus of Documents of Environmental Protection Commission of the State Council* (国务院环境保护委员会文件汇编), 263–279 (Volume 2, China Environmental Science Press, 1995).

31 Abram Chayes and Charlotte J. Kim, 'China and the United Nations Framework Convention on Climate Change', in Michael B. McElroy, Chris P. Nielsen, and Peter Lydon (eds), *Energizing China: Reconciling Environmental Protection and Economic Growth,* 515 (Harvard University Press, 1998).

32 Hatch (n 16), 51.

emissions included in the convention); developed countries had to provide financial assistance to developing countries (but no specific level of funding was mandated); and developed countries had an obligation to take "all practicable steps to promote, facilitate and finance, as appropriate, the transfer of . . . technologies"[33] (but China and other developing countries were unable to get a commitment from developed countries to transfer technology on "concessional and preferential terms").[34]

China saw itself already in this early phase as a major world power and wanted to be seen as a responsible and constructive international citizen,[35] but it did not play a true leading role in international climate negotiations. As discussed above, economic growth remained its top priority during this period, and its international negotiating positions were thus dominated by agencies with an economic focus. During the 1991 negotiations, China was unwilling to bear any obligation in reducing emissions. By stressing that economic development was a necessary precondition for developing countries to take on measures to address climate issues, China settled into the position that it was the responsibility of developed countries to reduce emissions first.

At the domestic level, from the late 1980s to the mid 1990s, the Chinese government had taken some measures to address the problems of environmental pollution. Some of these policy measures, although they were not climate-directed, did have the effect of limiting China's GHG emissions. They included afforestation, energy conservation, fuel switching, energy-price reform, and investment in energy-efficient technologies. The afforestation policy was originally used to help protect against soil erosion.[36]

China increased its forest area from 8.6% to 13% between 1949 and 1991.[37] However, less than 40% of the trees planted survived, due to China's emphasis on quantity over "quality".[38] When China gave high priority to reducing the energy

33 United Nations Framework Convention on Climate Change (UNFCCC), art. 4, para. 5 (opened for signature 9 May 1992, entered into force 21 March 1994), S. Treaty Doc. No. 102–138, 1771 U.N.T.S. 107 [UNFCCC].

34 Daniel Bodansky, 'The United Nations Framework Convention on Climate Change: A Commentary' (1993) 18(2) *Yale Journal of International Law* 451, 451–558.

35 Yufan Hao, 'Environmental Protection in Chinese Foreign Policy' (1992) 11(3) *Journal of Northeast Asian Studies* 25, 25–46; Robert D. Perlack, Milton Russell and Zhongmin Shen, 'Reducing Greenhouse Gas Emissions in China: Institutional, Legal and Cultural Constraints and Opportunities' (1993) 3(1) *Global Environmental Change* 78, 86; Alastair Iain Johnston, 'China and International Environmental Institutions: A Decision Rule Analysis', in Michael B. McElroy, Chris P. Nielsen and Peter Lydon (eds), *Energizing China: Reconciling Environmental Protection and Economic Growth,* 559, 563 (Harvard University Press, 1998); Michel Oksenberg and Elizabeth Economy, 'Introduction: China Joins the World', in Elizabeth Economy and Michel Oksenberg (eds), *China Joins the World: Progress and Prospects,* 21 (Council on Foreign Relations Press, 1999).

36 Hatch (n 16), 56.

37 Elisabetta Capannelli and Omkar L. Shrestha, 'Environmental Challenges in the People's Republic of China and Scope for Bank Assistance' (Asian Development Bank, December 1993) 3.

38 Vaclav Smil, *China's Environmental Crisis: An Enquiry into the Limits of National Development,* 59, 61 (Routledge, 1993).

intensity of its economy in its Eighth Five-Year Plan (1991–1995), energy conservation became a key policy measure to reduce energy intensity. However, China did not have enough funds to invest in energy conservation at that time and relied heavily on foreign investment in energy-efficient technologies.[39] Instead of investing in energy conservation, the Chinese government established energy conservation departments in government and in many large and medium-size enterprises, disseminated information on energy conservation, and provided incentive policies to reward energy conservation and penalize energy waste.[40] Notwithstanding the lack of investment, the energy intensity of China's economy still declined by 5.6% between 1980 and 1988.[41] But this decline was caused to a large extent by economic restructuring that preferred light over heavy industry.[42]

Fuel switching from coal to cleaner energy was also considered for energy conservation during this period. For example, the government considered reducing coal use for electricity generation by increasing investment in nuclear power and hydroelectric facilities.[43] However, because the costs of both alternatives were far more expensive than coal-burning thermal plants, they could not be promoted on a large scale.[44] Renewables, such as wind, solar, and biomass, as well as experimental geothermal and ocean energy, were still heavily reliant during this period on collaboration with foreign companies and institutions, and thus were slow to develop.[45] Another solution recommended by energy experts for energy conservation was energy-price reform. Under the centrally planned economy, coal prices in China were not determined by the market and thus had been relatively low for quite a long time. The low prices further impeded the switch to renewables. Energy-price reforms were initially implemented by the government in the mid 1980s, but they stalled in 1988 due to the resistance of production ministries relying on low-cost energy and that of political conservatives concerned about growing inflation and loss of influence.[46] It was not until the mid 1990s that coal prices came to be determined by the market.

39 An official in the Ministry of Energy indicated that the amount of funds spent on energy conservation depended on "foreign friends who would be willing to invest and provide technical assistance in China's energy conservation program to control China's contribution to global warming". See Gaye Christoffersen, 'China's "Comprehensive" Energy Policy', in Thomas W. Robinson (ed), *The Foreign Relations of China's Environmental Policy* (The American Enterprise Institute (Conference Report), 1992) 49.

40 Perlack, Russell, and Shen (n 35), 81.

41 Zongxin Wu and Zhihong Wei, 'Policies to Promote Energy Conservation in China' (1991) 19(10) *Energy Policy* 934, 937.

42 Catherine Locatelli, 'China's Energy Policy: Energy and Economic Development' (1989) 1(2) *Energy Studies Review* 144, 150; Smil (n 38), 126.

43 Geping Qu, 'China's Dual-Thrust Energy Strategy: Economic Development and Environmental Protection' (1992) 20(6) *Energy Policy* 500, 504.

44 It was estimated that the cost of nuclear power was four times more expensive than coal-generated electricity. There were only three nuclear power plants in operation as of November 1997. See Hatch (n 16), 56.

45 Hatch (n 16), 57.

46 Lester Ross, 'The Politics of Environmental Policy in the People's Republic of China' (1992) 20(4) *Policy Studies Journal* 628, 634.

During the early phase, there was no specific climate-oriented law or policy in China. Both its domestic policies and international negotiating stances, while having the effect of limiting China's GHG emissions to some extent, saw economic growth as its top priority. In this context, China was willing to address climate issues only in ways that did not undermine its economic development. As such, China played a back-seat role in international climate negotiations during this phase.

Phase II (1995–2007)

In 1995 and following, China gradually promoted climate change as a policy priority. An example is the creation of the high-level National Climate Change Countermeasures Coordination Group in 1998, which was led by the then-powerful State Development and Planning Commission (SDPC).[47] However, the institutionalization of SDPC's leading role in China's climate governance revealed that economic development was still China's top priority, since the primary task of the SDPC – similar to its predecessor, the SPC – was maximizing economic development. This implied that China's climate change negotiating position would be subject to the goal of economic development. Therefore, some contend that this institutional promotion of climate change was not an effort to enhance state capacity to address climate issues but rather a defensive move to ensure that climate change policy would not excessively impede economic growth.[48]

Many developed countries, but also, increasingly, vulnerable developing countries, were beginning to express dissatisfaction with China's refusal to commit to a GHG-emission-reduction target. In February 1995, during the final negotiation session of the Intergovernmental Negotiating Committee leading up to the First Conference of the Parties (COP 1) to the UNFCCC, China questioned the scientific conclusion of IPCC assessments that served as the basis for those arguing the inadequacy of existing commitments contained in the UNFCCC.[49] Meanwhile, it opposed proposals for a protocol containing binding commitments to reduce GHG emissions for all parties, contending that the parties should fully implement their commitments contained in the UNFCCC before taking further ambitious actions.[50] This position threatened the cohesion of the G-77, since the Alliance of Small Island States (AOSIS) was pushing for a protocol requiring developed countries to reduce their CO_2 emissions to 20% below 1990 levels by 2005.[51] To avoid a breakdown in negotiations, at COP 1, China compromised by acknowledging the inadequacy of developed countries' existing commitments and the

47 The State Development and Planning Commission (SDPC), before 1998 known as the State Planning Commission (SPC), was reorganized as the National Development and Reform Commission (NDRC) in 2003. See China.com.cn (中国网) (n 22).

48 Wang (n 8), 640.

49 Hatch (n 16), 52.

50 Chayes and Kim (n 31), 504.

51 Hatch (n 16), 53.

need to include quantified emission-limitation and emission-reduction objectives with specified time-frames in the proposed protocol.[52]

The Kyoto Protocol was signed at COP 3, in December 1997. Later, during COP 5 in Bonn in 1999, China held the following stance in the negotiations:[53]

1 Before becoming a middle-level developing country, China would not accept binding GHG-emission-limitation targets. However, the Chinese government would make a serious effort to dampen the increase in its GHG emissions. China would continue to promote and participate in international cooperation.
2 Developed countries should provide China with technology and financial assistance to enhance its ability to deal with climate change, in accordance with the UNFCCC.
3 The COP should address the following issues:

   i Encourage developed countries to fulfil their commitment to reduce GHG emissions and provide technology and financial assistance in accordance with UNFCCC Article 4.
   ii Encourage developed countries to expedite the process of ratification of the Kyoto Protocol.
   iii Add the right of a party to determine the measures taken to address climate change in its own specific circumstances.
   iv Discuss the principle of equity, including preventing the growing disparities in per capita GHG emissions and energy-consumption levels between developed and developing countries.

Two years later, at COP 7, in Marrakech, the Chinese government laid out its position as follows:[54]

1 Financial assistance and technology transfer are key for developing countries to enhance their ability to address climate change.
2 The environmental effect of the Kyoto Protocol depends on how carbon sinks work. A cautious attitude on the issue of carbon sinks is necessary.
3 The Kyoto Protocol's flexibility mechanisms are helpful for developed countries in reducing their emissions at low cost. The CDM will contribute to the sustainable development of developing countries. There should be an early establishment of the CDM Executive Board and a prompt launch of the CDM.
4 Strict compliance with the rules of the Kyoto Protocol is needed.

52 These compromises led to the 'Berlin Mandate' at COP1. See Chayes and Kim (n 31), 526–527.

53 Speech by Liu Jiang, Head of China's Delegation to the 5th Conference of the Parties to the UNFCCC in 1999 (18 July 2002) (in Chinese), translated by Zhang Haibin, 'China and International Climate Change Negotiations' (March 2013) WeltTrends Online-Dossier, 4 <http://welttrends.de/res/uploads/Zhang_China-and-International-climate-change-negotiations.pdf> accessed 25 January 2020.

54 Speech by Liu Jiang, Head of China's Delegation to the 7th Conference of the Parties to the UNFCCC in 2001 (18 July 2002) (in Chinese, translated by Zhang (n 53), 4).

During the second phase, China continued its resistance to a binding target to reduce GHG emissions. It also maintained its attitude on increased technology transfer and financial assistance from developed countries to developing countries. Moreover, it continued to insist on using per capita emissions as a criterion for deciding which countries should have such targets.

There were also some discontinuities in China's positions from the first to the second phase. In the first phase, China did not stress the need for developed countries to take a leading role in GHG-emission mitigation; however, during the second period, China clearly took that position. Another significant change was China's view on the Kyoto Protocol's flexibility mechanisms – emission trading, Joint Implementation (JI), and the Clean Development Mechanism (CDM). Although it initially hesitated to support the flexibility mechanisms in 1999,[55] China gradually came to support them, especially the CDM, in 2001.[56] But its support for this mechanism was not without reservation. China insisted that the CDM's Certified Emission Reductions be used only for complying with a portion of a developed country's target.[57] Moreover, it wanted the CDM projects to be bilateral and project-based, rather than fund-based with multilateral financing, since the latter was likely to weaken the negotiating leverage of developing countries (who tend to be under-represented in multilateral funds) and also likely to develop into a global emission-trading system (ETS), which might attempt to bring developing countries into premature commitments on emission caps.[58]

Domestically, during the second period, China committed itself to address climate change and associated environmental challenges through a series of actions. In 1997, China promulgated the Energy Conservation Law.[59] As part of China's energy policy, this law provided a framework for public–private partnerships (PPP) related to energy projects.[60] To promote energy efficiency, in 2004 a range of measures, including China's first energy-intensity reduction targets, were outlined by China in the Medium and Long Term Energy Conservation Plan

55 Chris P. Nielsen and Michael B. McElroy, 'Introduction and Overview', in Michael B. McElroy, Chris P. Nielsen, and Peter Lydon (eds), *Energizing China: Reconciling Environmental Protection and Economic Growth*, 39 (Harvard University Press, 1998).

56 Ibid.

57 Hatch (n 16), 54. Some contend that this concern was based on the belief that allowing the sale of CERs to third parties would increase the number of CDM projects. See Kristian Tangen, Gørild Heggelund, and Jørund Buen, 'China's Climate Change Positions: At a Turning Point?' (2001) 12(2&3) *Energy & Environment* 237, 242.

58 Xu Yugao, 'Different Perspectives between the North and the South on CDM: Viewpoints, Modalities, and Implications' (Development Research Academy for the 21st Century, Tsinghua University, Beijing, 1999) 10–11.

59 Energy Conservation Law of the People's Republic of China (中华人民共和国节约能源法), Standing Committee of the National People's Congress (adopted on 1 November 1997, amended on 28 October 2007 and 2 July 2016) (for an English version, see <www.lse.ac.uk/GranthamInstitute/wp-content/uploads/laws/2041.pdf> accessed 25 January 2020).

60 For more information on China's Energy Conservation Law and its relationship with the PPP-Energy projects, see <https://ppp.worldbank.org/public-private-partnership/sector/energy>.

for 2020.[61] In 2005, the Renewable Energy Law was passed. It established key policies, including national renewable-energy targets, a mandatory connection-and-purchase policy, a national feed-in-tariff system, and arrangements for the cost-sharing and funding of renewable-energy incentives.[62] Binding national targets on energy intensity and the reduction of major pollutants were introduced in the 11th Five-Year Plan (2006–2010).[63]

However, there was a discontinuity between China's domestic climate policy and its negotiating position in the international arena. In the context of an energy-intensive growth model for the country, policymakers were careful to avoid linking domestic action on climate change to any presumptions of international obligation, since emission targets could potentially impede economic development.[64] Although the Chinese government during this period laid the foundation of its climate policy, took measures to promote industrial efficiency, and reduced GDP energy intensity, it was unwilling to subject itself to any international climate regime that threatened to undermine its energy-intensive growth model.

Compared to the early phase, this phase saw China's shift to a more active role in the global climate change negotiations. During the negotiations leading to the 1997 Kyoto Protocol, China enhanced the cohesion of the G-77 by making compromises. As a major force representing developing countries in the climate negotiations, China strived for more benefits for developing countries on the basis of fairness and sovereignty. At the same time, it avoided any commitments and stressed the leading role of developed countries in addressing global climate issues. This position was in accordance with its focus on economic development. Although the Chinese government had taken climate issues into consideration to a greater extent than before, while developing its economic-development strategies, its efforts to mitigate GHG emissions were for the purpose of realizing those broader strategies. Thus, the majority of its climate efforts during this phase were in the form of policies implemented throughout the economy (particularly the energy sector) that had a side effect of reducing GHG emissions, rather than being climate policies per se.

61 NDRC, 'China Medium and Long Term Energy Conservation Plan' (25 November 2004) <fourfact.com/images/uploads/China_Energy_Saving_Plan.pdf> accessed 25 January 2020.

62 Renewable Energy Law of the People's Republic of China (中华人民共和国可再生能源法), Standing Committee of the National People's Congress (adopted on 28 February 2005, amended on 26 December 2009) (for an English version, see <http://english.mofcom.gov.cn/article/policyrelease/Businessregulations/201312/20131200432160.shtml> accessed 25 January 2020).

63 Guidelines of the Eleventh Five-Year Plan for National Economic and Social Development (中华人民共和国经济和社会发展第十一个五年计划) (for an English version, see <https://policy.asiapacificenergy.org/sites/default/files/11th%20Five-Year%20Plan%20%282006-2010%29%20for%20National%20Economic%20and%20Social%20Development%20%28EN%29.pdf> accessed 25 January 2020).

64 Isabel Hilton and Oliver Kerr, 'The Paris Agreement: China's "New Normal" role in international climate negotiations' (2017) 17(1) *Climate Policy* 48, 49.

Phase III (2007-Present)

China released its first National Climate Change Program[65] in 2007 and established the National Leading Group on Climate Change, led by the premier of the State Council (China's prime minister). Such a premier-led leading group marks a significant promotion of the policy priority of climate change in the Chinese government.

During the 2007 Bali COP, China's delegation to the global climate change negotiations took the following positions:[66]

1 Of utmost importance was to strengthen international cooperation on climate change and fully implement the UNFCCC and the Kyoto Protocol.
2 The UNFCCC and the Kyoto Protocol are the major legal frameworks through which the international community is to address climate change. They do not close the door to regional cooperation on climate change. Regional cooperation should function as a helpful complement to the UNFCCC and the Kyoto Protocol rather than replace or weaken them.
3 Fundamental to the implementation of the UNFCCC is a commitment to turning the mitigation, adaptation, and transfer of technology and finance under the Convention into actions.
4 Fundamental to the implementation of the Kyoto Protocol is a resolve by developed countries to reach their binding targets for the first commitment period and set new binding targets for the second commitment period.
5 China will agree to take measurable, reportable, and verifiable nationally appropriate mitigation commitments or actions on the condition that it is supported and enabled by technology, financing, and capacity building, in a measurable, reportable, and verifiable manner.

At the Copenhagen COP in 2009, China faced harsh criticism from developed countries over its perceived procrastination tactics and refusal to support binding emission-reduction targets. Although its negotiators pledged to reduce China's emission intensity by 40%–45% by 2020 from 2005 levels, they rejected any binding emission-reduction target that may constrain economic development.[67] Conrad notes the contradiction between the dynamism of China's domestic climate policy agenda and its tough position in the climate negotiations at Copenhagen, arguing that China's leadership was still hesitant to translate its domestic political interests

65 NDRC, 'China's National Climate Change Programme' (June 2007) <http://en.ndrc.gov.cn/newsrelease/200706/P020070604561191006823.pdf> accessed 25 January 2020.
66 Ibid; speech by Xie Zhenhua, Head of China's Delegation at the 13th Conference of the Parties to the UNFCCC (12 December 2007) (in Chinese, translated by Zhang (n 53), 5); UN, 'Report of the Conference of the Parties on its Thirteenth Session, held in Bali from 3 to 15 December 2007' (FCCC/CP/2007/6/Add.1, 14 March 2008).
67 Hilton and Kerr (n 64), 49.

into a coherent position on the international stage.[68] Hilton and Kerr maintain that climate change took second place to domestic economic development in China at the time of Copenhagen.[69]

During the UNFCCC's COP in Cancun, in 2010, China accepted that it would report biennially on its mitigation actions and emissions in biennial update reports.[70] The Chinese government's position at the Cancun COP was as follows:[71]

1 As a developing country, China was also a victim of global climate change; consequently, it strongly supported the needs of least-developed countries, African countries, and small island developing states.
2 Mitigation action would be taken by China to reduce CO_2 emissions per unit of GDP by 40%–45% by 2020 from 2005 levels. This would be an obligatory domestic target incorporated into China's medium- to long-term plan for national and economic development.
3 Developed countries should accomplish their ongoing commitments towards emission reductions and set new binding targets for the second commitment period of the Kyoto Protocol. The United States and any other developed countries not signing up to the second commitment period of the Kyoto Protocol should undertake comparable emission-reduction targets under the Convention, and developing countries should take effective voluntary adaptation actions according to their national conditions and capacity and join global efforts in tackling climate change.
4 Developed countries should provide finance, technology, and capacity-building support to developing countries through international cooperation. China would continue to provide assistance, to the extent of its ability, to other developing countries through South–South cooperation.

China has since become a most powerful player in climate negotiations. At the Paris COP in 2015, it took a constructive role in securing the new international agreement.[72] A key driver of the change between Copenhagen and Paris, as argued

68 Björn Conrad, 'China in Copenhagen: Reconciling the "Beijing Climate Revolution" and the "Copenhagen Climate Obstinacy"' (2012) 210 *The China Quarterly* 435, 435–455.
69 Hilton and Kerr (n 64), 50.
70 Jennifer Morgan and Deborah Seligsohn, 'What Cancun Means for China and the U.S.' (World Resources Institute, 16 December 2010).
71 Speech at the High Level Segment of COP16&CMP6 Delivered by Vice-Chairman Xie Zhenhua, National Development and Reform Commission, People's Republic of China, Cancun, 8 December 2010 (in Chinese, translated by Zhang (n 53), 5).
72 Qi Ye and Tong Wu, 'China's "yes" to New Role in Climate Battle' (Brookings Institution, 4 December 2015) <www.brookings.edu/articles/chinas-yes-to-new-role-in-climate-battle/> accessed 25 January 2020; Associated Press, 'China Played a Key Role in Global Climate Pact' (SFGATE, 13 December 2015) <www.sfgate.com/world/article/China-played-key-role-in-global-climate-pact-6695812.php> accessed 25 January 2020; Anthony H. F. Li, 'Hopes of Limiting Global Warming? China and the Paris Agreement on Climate Change' (2016) 1 *China Perspectives* 49, 49–54.

by Hilton and Kerr, was China's shift to the New Normal model of economic development[73] since the beginning of the Twelfth Five-Year Plan period (2011–2015), which, as indicated earlier, emphasized a transition from energy-intensive growth, which was based on heavy industry and exports, to a more balanced economy characterized by slower growth, an increasing role for services and domestic consumption, and a focus on innovation and low-carbon technologies.[74] The Twelfth Five-Year Plan set energy- and carbon-intensity targets, encouraged investment in low-carbon industries, increased R&D spending on low-carbon technologies, launched pilot carbon-trading schemes, and took steps to promote renewable energy so as to reduce reliance on coal in the energy mix.[75]

These measures were considered a radical change in economic policy[76] and were explicitly linked to climate policy.[77] In the context of this new model of economic development, environmental sustainability was prioritized and closely connected with "accelerating economic restructuring", no longer being seen as contradictory to economic policy.[78] The alignment between China's domestic economic and climate policies provided China with more flexibility during this period in its international climate policy and global climate change negotiations. In the New Normal era of development, emission reductions were no longer considered a burden to economic development but a strategic opportunity that would allow China to resolve its increasing domestic environmental crisis, improve its international image in the field of climate change, and promote the development of low-carbon technologies that would be urgently needed in the future. In its Nationally Determined Contribution (NDC) under the Paris Agreement, China committed to peak its CO_2 emissions by 2030, increase the non-fossil share of

73 The slogan "New Normal" was put forward by President Xi Jinping in 2014 in the context of China's economic downturn, which hit in 2011. National Bureau of Statistics of China, online statistics database <www.stats.gov.cn/english/Statisticaldata/AnnualData/> accessed 25 January 2020. For decades, China's economy had been dominated by double-digit economic growth, also referred to as "old normal growth". Old normal growth had faced challenges from energy shortage and environmental degradation. Under the pressure of flagging GDP growth after 2011, the government's focus shifted towards higher-quality growth, and China's domestic climate policy steadily increased in scope and ambition.

74 Hilton and Kerr (n 64), 48–58.

75 People's Republic of China (PRC), Twelfth Five-Year Plan (2011–2015) for National Economic and Social Development (中华人民共和国国民经济和社会发展第十二个五年规划纲要) (for English version, see <https://policy.asiapacificenergy.org/sites/default/files/12th%20Five-Year%20Plan%20%282011-2015%29%20for%20National%20Economic%20and%20Social%20Development%20%28EN%29.pdf> accessed 25 January 2020); Zhongxiang Zhang, 'Climate Mitigation Policy in China' (2015) 15(S1) *Climate Policy* S1, S1–S6.

76 Nicholas H. Stern, 'Raising Consumption, Maintaining Growth and Reducing Emissions: The Objectives and Challenges of China's Radical Change in Strategy and its Implications for the World Economy' (2011) 12(4) *World Economics* 13, 13–34.

77 State Council of the People's Republic of China, Energy Development Strategic Action Plan (2014–2020) (能源发展战略行动计划（2014–2020年）) <www.gov.cn/zhengce/content/2014-11/19/content_9222.htm> accessed 25 January 2020.

78 Ibid.

total primary energy consumption to 20% by 2030, and reduce the carbon intensity of GDP to 60%–65% below the 2005 level by 2030.[79]

China's strategic framework for climate change following the Paris COP is further supported by the Thirteenth Five-Year Plan (2016–2020), released in March 2016, which sets clear targets on reducing energy and carbon intensity, peaking carbon emissions, increasing energy production from renewable energy, eliminating surplus coal capacity, and developing green infrastructure during the Thirteenth Five-Year Plan period.[80] They include a target to reduce energy intensity per unit of GDP by 15% and reduce carbon emissions per unit of GDP by 18%.[81] These plans' targets are considered to be consistent with China's 2030 NDC targets, and given its current national policies, China is on track to overfulfil its 2030 NDC targets.[82]

To integrate its climate change targets and low-carbon development plans for specific sectors into its five-year plans, the Chinese government, in 2014, adopted the National Plan on Climate Change (2014–2020).[83] To improve energy efficiency, the NDRC and other ministries jointly formulated the Implementation for the Energy Efficiency Top-Runner System in December 2014, which involves setting stringent energy-efficiency standards for vehicles, buildings, appliances, and industrial equipment.[84] In 2016, the National Energy Commission suspended the approval of new coal mine projects through to 2019.[85]

Another sign of progress at the domestic level was the launch in 2013 of pilot ETSs in Beijing, Shanghai, Tianjin, Chongqing, Guangdong, Hubei, and Shenzhen, which was scaled up to a national ETS for power generation in 2017.

79 CAT, 'China – Pledges and Targets' (Climate Action Tracker, 2 December 2019) <https://climateactiontracker.org/countries/china/pledges-and-targets/> accessed 25 January 2020.

80 The Thirteenth Five-Year Plan for Economic and Social Development of the People's Republic of China (2016–2020) (中华人民共和国国民经济和社会发展第十三个五年（2016–2020）规划纲要) (for an English version, see <www.greengrowthknowledge.org/sites/default/files/downloads/policy-database/CHINA%29%20The%2013th%20Five-Year%20Plan%20%282016-2020%29.pdf> accessed 25 January 2020).

81 Ibid.

82 CAT, 'China – Country Summary' (Climate Action Tracker, 2 December 2019) <https://climateactiontracker.org/countries/china/> accessed 25 January 2020.

83 NDRC, 'National Plan on Climate Change (2014–2020) (国家应对气候变化规划（2014–2020年））' (September 2014) <https://policy.asiapacificenergy.org/sites/default/files/%E5%9B%BD%E5%AE%B6%E5%BA%94%E5%AF%B9%E6%B0%94%E5%80%99%E5%8F%98%E5%8C%96%E8%A7%84%E5%88%92%EF%BC%882014-2020%E5%B9%B4.pdf> accessed 25 January 2020.

84 Alina Averchenkova et al., 'Climate Policy in China, the European Union and the United States: Main Drivers and Prospects for the Future' (Grantham Research Institute on Climate Change and the Environment, Centre for Climate Change Economics and Policy, Bruegel, 17 November 2016) <https://bruegel.org/wp-content/uploads/2016/12/Averchenkova-et-al-2016.pdf> accessed 25 January 2020; Energy Charter Secretariat, 'China Energy Efficiency Report: Protocol on Energy Efficiency and Environmental Aspects' (Energy Charter Secretariat, 2018) 87.

85 'National Energy Commission: Coalmine Approvals Suspended in the Next Three Years' (国家能源局：未来三年暂停煤矿审批) *anychem.com* (30 December 2015) <http://coalchem.anychem.com/2015/12/30-16899.html> accessed 25 January 2020.

According to the Development Plan for a National Carbon Trading Market (Power Generation Sector), the development of the national ETS comprises three stages: a nationwide data-reporting, registration, and transaction-log system in the first year; piloting the transaction of allowances in the second year; and expanding the market to other sectors in the future, with the ultimate goal of establishing a carbon market with smooth circulation, effective regulation, fairness, and transparency.[86]

In addition to the proliferation of discrete energy and climate-related plans, policies, and projects during this phase, the Chinese government has, since 2010, been considering developing a framework law on climate change and energy transition.[87] The first draft of the Law on Combating Climate Change was released for public comment in March 2012.[88] The drafting of this law was still in progress at the time of writing and had not been put on the legislature's agenda.[89] China has laws, such as the Renewable Energy Law and the Energy Conservation Law, that can be used incidentally for mitigating GHG emissions or adapting to climate change.[90]

As noted, China's attitude towards participating in the international climate change regime has become increasingly flexible and cooperative. It has changed its attitude towards technology transfer and financial assistance. Compared to negotiations during earlier periods that advocated for the unilateral transfer of technology and financial assistance from developed countries, China is now more willing to pursue reciprocal technology cooperation in a South–South context. This attitude is consistent with the objective of Article 10 of the Paris Agreement, which calls generally for strengthening cooperative action on climate-technology development and transfer. Between 2006 and 2011, China financed and built the Kamchay Dam, the first large hydropower dam in Cambodia. Urban argues that in building this dam, Chinese dam builders transferred advanced hydropower technology to Cambodian recipients, but the knowledge transfer directed at strengthening indigenous innovation capabilities was limited.[91]

In terms of bilateral cooperation, China and the EU, in 2014, initiated a cooperation project on emission trading to support China's pilot ETSs and help China

86 Cheng Xiaofeng, Hu Ke, and Jiang Xinyan, 'China', in Theodore L. Garrett (ed), *The Environment and Climate Change Law Review,* 42, 51 (3rd edn, The Law Reviews, 2019).

87 Alina Averchenkova, 'Legislating for a Low Carbon and Climate Resilient Transition: Learning from International Experiences' (Real Instituto Elcano, Elcano Policy Paper, January 2019) 60.

88 'Law on Combating Climate Change of the People's Republic of China' (Draft for Public Comments) (中华人民共和国气候变化应对法》（征求意见稿）) (18 March 2012) <http://news.china.com.cn/txt/2012-03/18/content_24923504.htm> accessed 25 January 2020.

89 Tian Danyu, 'Promote Legislation on Combating Climate Change, Form a Synergy of Systematic Legal Systems' (推动应对气候变化立法 形成系统法律制度合力), *China Environment News* (中国环境报) (25 October 2019) 3 <http://epaper.cenews.com.cn/html/1/2019-10/25/03B/2019102503B_pdf.pdf> accessed 25 January 2020.

90 For more information on Chinese laws related to climate change, see Chapter 1 of this book.

91 Frauke Urban, 'China's Rise: Challenging the North-South Technology Transfer Paradigm for Climate Change Mitigation and Low Carbon Energy' (2018) 113 *Energy Policy* 320, 323–324.

to implement and develop its national ETS.[92] This project was extended until September 2020, with a €10 million contribution from the EU.[93] In addition to focusing on capacity and training activities, this bilateral cooperation project aimed to establish regular bilateral policy dialogues on ETS matters.[94] China's cooperative attitude in this respect corresponds with Article 6 of the Paris Agreement, which encourages bilateral or multilateral cooperation in a bottom-up, decentralized manner.

Outlook

Ever since China's entry into the global climate change negotiations at the end of the 1980s, there have been both continuities and discontinuities in its position and policies. During the early stage, discontinuities between global and domestic climate law and policy were apparent: the Chinese government focused on GDP growth, which relied on energy-intensive heavy industry, and was not willing to slow down its economic growth for mitigation purposes. After realizing that its energy-intensive growth model was unsustainable, tending to undermine economic development, the Chinese government started to shift to the New Normal model of development. Such a transition gradually bridged the gap between the international climate law and policy and China's domestic climate law and policy, which helped seal the Paris Agreement in 2015.

Although the resistance to a binding GHG-emission-reduction target remains unchanged, China's attitude towards the global climate change negotiations has become more flexible and cooperative. A series of key factors have been identified by scholars, both domestically and internationally, to explain the continuity and discontinuity in China's global and domestic climate change policy, among which national interest (such as abatement costs and ecological vulnerability) and the principle of equity are crucial in determining China's position in the climate change negotiations.[95]

92 European Commission, 'Emissions Trading: European Commission and China Hold First Policy Dialogue' (European Commission, 26 April 2018) <https://ec.europa.eu/clima/news/emissions-trading-european-commission-and-china-hold-first-policy-dialogue_en> accessed 25 January 2020.

93 Ibid.

94 Ibid.

95 See, e.g., Detlef Sprinz and Tapani Vaahtoranta, 'The Interest-Based Explanation of International Environmental Policy' (1994) 48(1) *International Organization* 77, 77–105 (arguing that a country's position and policies are determined by the ecological vulnerability of the country and the abatement cost borne by the country); Robert D. Putnam, 'Diplomacy and Domestic Politics: The Logic of Two-Level Games' (1998) 42(3) *International Organization* 427, 427–460 (arguing that a country's position in the global climate negotiations can be best explained by looking at the interactions between domestic and international politics); and Zhang (n 53) (arguing that the more equitable responsibility each country is willing to accept, the more likely it is that China will commit to reducing its GHG emissions and take a more cooperative attitude towards the global climate change negotiations).

From a historical perspective, the trajectory of China's participation in the international negotiations reveals that the principle of equity (which overlaps with the principle of common but differentiated responsibilities) has had a significant impact on China's willingness to share responsibility for combating climate change, and it is still likely to impede progress in global cooperation in the field of climate change. Technological and financial cooperation, as well as a growing sense of vulnerability to climate impacts, may provide important incentives to shift China's current position in a more positive direction. Although environmental concern has been elevated during the past decades, economic interests and the primacy of economic growth continue to play important roles in guiding China's climate law and policy. Compared to uncertain future costs, which may be potentially even higher, short-term costs tend to carry more weight in decision-making. Nonetheless, the increasing awareness of, and priority accorded to, climate change and the country's ambitious energy-efficiency goals promise a further reduction in carbon intensity in China.

4 Renewable energy laws and policies in China in the context of climate change

Mandy Meng Fang

Introduction

Surpassing the United States as the world's largest carbon emitter in 2006, China's energy consumption has increased rapidly in recent decades due to its unprecedented industrialization and urbanization, which shows no sign of slowing down.[1] The heavy reliance on coal, which accounted for around 60% of China's primary energy consumption in 2018 and cannot be substantially changed in the near future, has rendered the control of carbon emissions a challenging task for the country.[2] The trade-off between short-term economic benefits and long-term sustainable development presents a difficult challenge for the Chinese government.

There are three primary reasons for the increasingly widespread acknowledgement of the need to boost renewable-energy development in China. First, with mounting pressure to tackle environmental problems associated with the burning of fossil fuels, such as air pollution and global warming, renewable energy has emerged as a desirable alternative to fossil fuels. The public outcry in recent years against smog has given much urgency to the need for a cleaner growth path in China. It is also indisputable that China needs to be central to any serious global effort to address climate change. Second, the deployment and development of renewable energy can contribute to economic growth, since it brings opportunities for creating domestic jobs and increasing export growth. China and a rising number of other countries around the world have attached strategic importance to enhancing their competitiveness in the renewable-energy sector. Third, the gigantic energy demand and heavy reliance on imported fuels in China make energy security a significant issue that cannot be underestimated. Meeting energy

1 See 'BP Statistical Review of World Energy' (BP, June 2018) <www.bp.com/content/dam/bp/business-sites/en/global/corporate/pdfs/energy-economics/statistical-review/bp-stats-review-2018-full-report.pdf> accessed 12 August 2019.

2 Feng Hao and Tom Baxter, 'China's Coal Consumption on the Rise', China Dialogue (Beijing, 1 March 2019) <www.chinadialogue.net/article/show/single/en/11107-China-s-coal-consumption-on-the-rise> accessed 12 August 2019.

demand with renewable-energy resources will alleviate China's dependence on foreign energy resources and, thus, its vulnerability to energy disruptions.

In this chapter, the focus is on lawmaking and policymaking by the Chinese central government, which is the key decision-making body in the country's renewable-energy sector. The aim is to critically review the Chinese legal and policy framework for renewable energy and to explore how it could be improved to facilitate the transition to renewable energy and to mitigate climate change.

China has invested great efforts in developing renewable energy since the beginning of this century; even before that, a number of small-scale renewable energy programmes with the primary aim of addressing energy poverty in the rural regions were established in the mid 1990s.[3] Particularly in the wind and solar photovoltaic (PV) energy sectors, remarkable achievements have been made with respect to both manufacturing and installed capacity.[4] The rapid growth of wind and solar energy not only meets ever-increasing energy demand but also reduces carbon emissions that otherwise would have been generated by the burning of fossil fuels. The evolution of the policymaking and implementation processes over the past two decades in China's renewable-energy sector is of paramount research interest. Hence, the aim of this chapter is to analyse those developments.

The next section discusses the rationale for promoting renewable energy in China, consisting of environmental, economic, and energy-security reasons. Although addressing climate change was not the primary motivation for the government's renewable-energy development strategy, its importance has been increasingly recognized and factored into renewable-energy policymaking. A subsequent section details the laws and policies that have been promulgated to incentivize renewable-energy development in China in recent decades. The focus is on central-level laws and policies, which is not to downplay the fact that local governments have been increasingly active in designing and implementing rules and policies to facilitate renewable-energy development. The astonishing increase in wind and solar energy in China over the past two decades makes these two sectors an intriguing focus for an analysis of the evolution of laws and policies.

The rationale for promoting renewable energy in China

China's government has placed strategic priority on the development of renewable energy for a multitude of reasons, which merits an in-depth analysis. This section identifies three primary driving forces that have motivated China's government to shape an enabling environment for the rapid growth of renewable energy: environmental, economic, and energy-security reasons.

3 Jinglin Fan and others, 'The Development of China's Renewable Energy Policy and Implications for Africa' (2018) 394(4) *IOP Conference Series Materials Science and Engineering*, 2.

4 Christopher Dent, 'China's Renewable Energy Development: Policy, Industry and Business Perspectives' (2015) 21(1) *Asia Pacific Business Review* 26.

Environmental reasons

Due to its long-standing heavy reliance on the use of coal to meet energy demands and promote economic growth, China has faced acute environmental problems, such as air pollution, which have caused adverse economic and health consequences.[5] The rising public awareness of air pollution, as reflected in a number of high-profile mass protests, has pushed the need to develop renewable energy as a replacement for coal to the forefront.[6] The reduction or phase-out of carbon-intensive economic activities forms an essential part of China's environmental action to solve the problem of air pollution. Put differently, reducing the use of fossil fuels, particularly coal, is not only an environmental issue but an important aim for domestic social stability in China.

Another environmental reason is the imperative faced by China to mitigate climate change, which is a worldwide struggle that cannot be advanced without China's participation. In the commitment made by China in its Nationally Determined Contribution to the 2015 Paris Agreement, the country is to peak its carbon emissions around 2030 while making best efforts to peak early; it is also to achieve 20% non-fossil energy as a proportion of primary energy supply by 2030.[7] In practice, addressing climate change has gradually become an integral part of China's renewable-energy strategy; and the ability of renewables to produce cleaner energy renders them a significant part of the solution to the pressing environmental problems that China needs to address.

Economic reasons

The development of renewable energy and the transition to a low-carbon economy can deliver substantial benefits in economic terms, such as export growth, employment opportunities, and tax revenue.[8] The manufacturing of renewable-energy equipment, such as wind turbines and solar panels, has been designated as a strategic and pillar industry in China due to its huge potential in increasing economic and technological competitiveness.[9] China can make use of its sizeable domestic market to achieve scale economies in a rapid manner.[10]

5 Dominic Chiu, 'The East Is Green: China's Global Leadership in Renewable Energy' (6 October 2017, Center for Strategic and International Studies) <www.csis.org/east-green-chinas-global-leadership-renewable-energy> accessed 12 August 2019.

6 Ibid.

7 See National Development and Reform Commission (NDRC), 'Enhanced Actions on Climate Change' <http://www4.unfccc.int/ndcregistry/PublishedDocuments/China%20First/China%27s%20First%20NDC% 20Submission.pdf> accessed 12 August 2019.

8 For a comprehensive discussion of the economic benefits that renewable energy development can bring, see IRENA, 'Renewable Energy Benefits: Measuring the Economics' (IRENA, 2016).

9 Kevin Lo, 'A Critical Review of China's Rapidly Developing Renewable Energy and Energy Efficiency Polices' (2014) 29 *Renewable and Sustainable Energy Reviews* 508, 509.

10 Sufang Zhang and others, 'Interactions between Renewable Energy Policy and Renewable Energy Industrial Policy: A Critical Analysis of China's Policy Approach to Renewable Energy' (2013) 62 *Energy Policy* 342, 351.

Energy-security reasons

Serious energy-security challenges, such as energy supply risk and energy price fluctuation, have become acute problems in China. Since the global financial crisis of 2007–08, China has increased its imports of energy resources, and half of the oil that China consumes comes from foreign sources, especially from countries in the Middle East, Africa, and Central Asia.[11] China has shifted from being an oil-exporting country in the early 1990s to one of the largest oil-importing countries in the world, with an oil-import dependency rate of more than 50%.[12] Facing high growth in primary energy demand as well as high energy intensity, China's reliance on foreign energy resources shows no sign of abating in the foreseeable future.[13] The volatile international prices for oil, gas, and coal have caused renewed concern about energy security in China.[14]

As an inexhaustible resource, renewable energy can provide a growing and vital contribution to energy security in two ways. First, the deployment and development of renewable energy in China can meaningfully reduce the country's reliance on foreign energy sources and alleviate its vulnerability to price spikes. Second, China enjoys abundant renewable-energy resources that can be rapidly and efficiently developed. Therefore, facing an urgent need to alleviate dependence on foreign energy supply, China will need to develop renewable energy.

China's legal and policy framework for the development of renewable energy

Laws and policies on renewable energy are essential to facilitate renewable-energy development.[15] China's legal and policy framework for renewable energy has evolved over two decades. This section will present China's renewable-energy laws and policies with a focus on those adopted at the national level to support the development and deployment of wind and solar PV energy – the two most rapidly growing renewable-energy resources in China.[16] An examination of the laws and policies designed to develop wind and solar energy in China sheds light on the role of lawmaking and policymaking in China's renewable-energy development.

11 Zhang Jian, 'China's Energy Security: Prospects, Challenges, and Opportunities' (Brookings Institution, July 2011) <www.brookings.edu/wp-content/uploads/2016/06/07_china_energy_zhang_paper.pdf> accessed 12 August 2019, 2.

12 Guy Leung, 'China's Energy Security: Perception and Reality' (2011) 39(3) *Energy Policy* 1330, 1331.

13 Dent (n 4), 33.

14 Zhang (n 11), 3.

15 Ming Zeng, Chen Li, and Lisha Zhou, 'Progress and Prospective on Policy System of Renewable Energy in China' (2013) 20 *Renewable and Sustainable Energy Reviews* 36, 42.

16 IRENA, 'Renewable Energy Prospects: China' (November 2014) <www.irena.org/publications/2014/Nov/Renewable-Energy-Prospects-China> accessed 12 August 2019.

China's early laws on renewable energy

The development of the legislative process for renewable energy in China could be dated back to the 1980s, when the Agriculture Law stated that the use of straw from agricultural production could provide a new source of energy and help alleviate rural energy shortage.[17] Article 16 of the Water Law, enacted in 1988, encouraged the development and use of hydropower resources, aiming to meet national economic-development needs and people's livelihoods.[18] In 1996, the Electric Power Law entered into force. Article 5 of this law provides that "the State encourages and supports the use of renewable energy and clean energy for electricity generation".[19] The language in this law – using a verbal tense that is typical of much Chinese statutory drafting – is "soft", which brings its enforcement into doubt. The Energy Conservation Law, enacted in 1997, recognized the need to encourage and support the development and use of new energy resources and renewable-energy resources.[20] That law's language is also vague and non-directive.

This brief review of the early stage of legal development indicates that although references to renewable-energy use and development can be found in several statutes, they are fragmented, lack continuity, use weak language, and do not create long-term expectations about growth in the sector.[21] The absence of a basic and dominant statute to regulate the development of renewable energy in China was the main problem during this period. The laws and policies adopted at the end of twentieth century that pertained to renewable energy did not provide sufficient incentives for its deployment and development.

Landmark development: the passage of the renewable energy law

The Renewable Energy Law of 2005[22] was China's first law on renewable energy, providing legislative support to its exploitation and use.[23] The law established the basic legal system and policy framework for China's development of renewable

17 See Agriculture Law of the People's Republic of China (中华人民共和国农业法), Standing Committee of the National People's Congress (adopted on 2 July 1993).

18 See Water Law of the People's Republic of China (中华人民共和国水法), Standing Committee of the National People's Congress (adopted on 21 January 1988).

19 See Electric Power Law of the People's Republic of China (中华人民共和国电力法), Standing Committee of the National People's Congress (adopted on 28 December 1995).

20 See Energy Conservation Law of the People's Republic of China (中华人民共和国节约能源法), Standing Committee of the National People's Congress (adopted on 11 November 1997).

21 Cao Wei and Li Yanfang, 'Framework of Laws and Policies on Renewable Energy and Relevant Systems in China under the Background of Climate Change' (2012) 13(4) *Vermont Journal of Environmental Law* 823, 833.

22 Renewable Energy Law of the People's Republic of China (中华人民共和国可再生能源法), Standing Committee of the National People's Congress (adopted on 28 February 2005, amended on 26 December 2009).

23 Joel Eisen, 'China's Renewable Energy Law: A Platform for Green Leadership?' (2010) 35(1) *William and Mary Environmental Law and Policy Review* 1, 25.

energy.[24] The goals of the law include increasing the domestic energy supply, optimizing the energy structure, ensuring energy security, protecting the environment, and realizing the sustainable development of China's economy and society.[25] The following key mechanisms were established under the law to promote the growth of renewable energy:

1. A national renewable-energy target, as well as central and local planning for renewable-energy development and use.[26]
2. A mandatory connection-and-purchase policy by which grid companies are required to sign an agreement with renewable-electricity generators to purchase the electricity generated from the generators in the full amount and provide associated grid-connection services.[27]
3. An on-grid electricity price for renewables, which pays renewable-electricity generators a fixed amount for each kilowatt hour of electricity generated over and above the wholesale electricity price for desulphurized coal-fired power.[28]
4. A cost-sharing mechanism funded through a surcharge on electricity sales and a Renewable Energy Development Fund.[29]

The Renewable Energy Law was silent on many important details, which needed to be added through rules and regulations.[30] Soon after the passage of the law, for instance, the National Development and Reform Commission (NDRC) issued the Provisional Administrative Measures on Pricing and Cost Sharing for Renewable Energy Power Generation;[31] the Provisional Regulation on Renewable Energy Surcharge Balancing;[32] the Regulation on the Administration of Power Generation

24 Li and Cao (n 21), 833.
25 Renewable Energy Law 2005, art. 1.
26 Ibid., arts. 7 and 8.
27 Ibid., art. 14.
28 Ibid., art. 19.
29 Ibid., arts. 20 and 24.
30 Barbara Finamore, 'China Renews Its Commitment to Renewable Energy' (1 February 2010, Natural Resources Defense Council) <www.nrdc.org/experts/barbara-finamore/china-renews-its-commitment-renewable-energy> accessed 12 August 2019; Jack Su, Simone Hui, and Kevin Tsen, 'China Rationalizes Its Renewable Energy Policy' (2010) 23(3) *Electricity Journal* 26, 28.
31 Notice of NDRC on Issuing Provisional Administrative Measures on Pricing and Cost Sharing for Renewable Energy Power Generation (国家发展改革委关于印发可再生能源发电价格和费用分摊管理试行办法的通知), NDRC Order No. (2006)7, 20 January 2006. It sets out the principles for renewable-energy power pricing and cost sharing. In particular, it identifies the level of wind and biomass power pricing and clarifies all costs related to renewable-energy power that will be covered by the renewable-energy surcharge.
32 Notice of NDRC on Issuing Provisional Regulation on Renewable Energy Surcharge Balancing (国家发展改革委关于印发可再生能源电价附加收入调配暂行办法), NDRC Order No. (2007)44, 11 January 2007. It identifies the procedure for provincial power utilities to collect the renewable-energy surcharge, the method for allocating this revenue among the provinces, and the role of the monitoring body in this process.

from Renewable Energy;[33] the Guiding Catalogue for Development of the Renewable Energy Industry;[34] and the Provisional Administrative Measures on the Renewable Energy Development Fund.[35]

In the first few years following the passage of the law, China's renewable-energy deployment grew strongly. The total installed capacity for wind power rose to 2.6 GW in 2006, new installations in that year alone exceeding the combined total for the past two decades.[36] However, the regulation relating to grid connection, integration, use, and subsidy allocation was not implemented in a satisfactory manner, with severe implications for the sustained development of renewable energy in China. In particular, the requirement imposed on grid companies to connect and purchase all electricity from renewable sources suffered from a lack of monitoring and compliance requirements.[37] The target setting emphasized the installed capacity more than the actual use of electricity, which led to a high level of renewable-energy curtailment. Payments to renewable electricity generators for on-grid electricity supply were frequently delayed. The deficiencies associated with the design and implementation of the Renewable Energy Law led to calls for reform.[38]

In 2009, the Renewable Energy Law was amended. Three major amendments are of interest. The first mandated more-detailed planning and coordination, including coordination of renewable-energy development plans at the local level with plans at the national level so as to ensure that local governments act in accordance with the national plan.[39] Greater coordination was also mandated between power-sec-

33 Notice of NDRC on Regulation on the Administration of Power Generation from Renewable Energy (国家发展改革委关于印发可再生能源发电有关管理规定), NDRC Order No. (2006)13, 5 January 2006. It sets out approval procedures for renewable-energy projects and further identifies the responsibilities of utilities and power generators. It obliges utilities to allow renewable-energy facilities to connect to the grid.

34 Notice of NDRC on Guiding Catalogue for Development of the Renewable Energy Industry (国家发展改革委关于印发可再生能源产业发展指导目录的通知), NDRC Order No. (2005)2517, 29 November 2005. It identifies the renewable-energy technologies that will be supported by the government and identifies the economic policy instruments that will apply to them.

35 Notice of MOF on Provisional Administrative Measures on the Renewable Energy Development Fund (财政部关于印发可再生能源发展专项资金管理暂行办法的通知), MOF Order No. (2006)237, 30 May 2006. It sets out the criteria for the use of the Renewable Energy Development Fund, identifies "priority areas", and provides application and approval procedures.

36 Zhongxiang Zhang, 'China in the Transition to a Low-Carbon Economy' (2010) 38 *Energy Policy* 6638, 6642.

37 IRENA, 'Renewable Energy Prospects: China, RE Map 2030 Analysis' (November 2014) <www.irena.org/-/media/Files/IRENA/Agency/Publication/2014/IRENA_REmap_China_report_2014.pdf> accessed 12 August 2019, 36.

38 'China Rationalizes Its Renewable Energy Policy' (Mayer Brown, 2 August 2010) <www.mayerbrown.com/en/perspectives-events/publications/2010/08/china-rationalizes-its-renewable-energy-policy> accessed 12 August 2019.

39 The 2005 Renewable Energy Law required only the provincial-level energy departments to formulate a renewables plan for their own jurisdictions without seeking direction or oversight at the national level. The 2009 Law enables the central government to access, oversee, and better coordinate the details of provincial-level renewables planning with grid-development plans.

tor and transmission planning so as to reduce renewable energy curtailment. The second major amendment of interest was aimed at improving the implementation and enforcement of the full-electricity-purchase and grid-connection obligations. The measures introduced to achieve that objective included a renewable-energy quota system that set generation-based targets for grid companies;[40] a priority-dispatch system favouring renewable electricity;[41] and technical standards for grid interconnections.[42] The amendments also added deadlines and economic penalties for power companies that failed to comply with the guaranteed-purchase requirement. The third major amendment was to streamline the Renewable Energy Development Fund to remove delays in the payment of fiscal incentives to renewable-energy generators.[43]

With the momentum added by the amended law, China's investment in renewable energy in 2009 reached $34.6 billion, surpassing that of the United States.[44] Nevertheless, the amended law was by no means a silver bullet to solve all of China's problems in renewable-energy development. A number of issues have since emerged that have threatened the sustained development of renewable energy in China. In the following sections, wind and solar energy are used as two case studies through which to analyse the current deficiencies.

Overview of developments in law and policy in China's wind-energy sector

China's onshore wind-power resources at an altitude of 10 metres amount to 3,226 GW for electricity, of which 253 GW is available for development and use.[45] Wind-power resources in coastal waters total around 750 GW.[46] The developmental potential is therefore large.

At an early stage of development, wind power in China received little government support.[47] Then, China's State Planning Commission set the first wind-energy target, which was to develop 1,000 MW of installed capacity by 2001 (it was known as the Ride the Wind Programme).[48] In the early 2000s, the government introduced a number of incentives to boost wind-energy development,

40 See Renewable Energy Law 2009, art. 14.

41 Ibid.

42 Ibid., art. 11.

43 The Renewable Energy Law 2009 centralizes and streamlines the flow of funds by requiring that surcharges be pooled together in the new Renewable Energy Development Fund, from which grid companies can seek compensation for their additional costs of purchasing electricity generated from renewable-energy sources as well as those incurred from grid connection.

44 Zhang (n 36), 6642.

45 Li Junfeng et al., *China Wind Power Report* (China Environmental Science Press, 2007), 2.

46 Ibid.

47 Ibid., 5.

48 Hanjie Wang et al., 'Wind Power in China: A Cautionary Tale' (International Institute for Sustainable Development, 2016) <www.iisd.org/sites/default/files/publications/wind-power-in-china-cautionary-tale.pdf> accessed 12 August 2019.

including a Wind Farm Concession Programme initiated by the NDRC in 2003.[49] Under that programme, investors and developers of wind-power projects larger than 50 MW were selected through bidding, with the aim of solving issues such as grid connection, tariffs, and cost distribution.[50] Most of the bidding was for large wind farms in China's wind-abundant north.[51] The bidding-based programme introduced competition to both the management and construction of wind farms, potentially reducing costs. However, as noted by Zhang, one of the notable deficiencies of the programme was that investors underbid to win a project, which could endanger the quality and viability of the project.[52] The practice of bidding at an unrealistically low price favours large, incumbent, wind-power developers, particularly state-owned enterprises (SOEs) free of profit-seeking motives, driving out independent power developers.[53]

The Chinese government was also determined to boost domestic manufacturing capacity in the wind sector. Because domestic manufacturers of wind-power equipment hardly existed at that stage, localization requirements are found in a variety of supportive schemes for wind power, aiming to promote the domestication of the production of large-scale wind turbines and improve manufacturing competitiveness.[54] By the end of 2004, the market share of locally manufactured wind turbines had reached 18%.[55] Installed capacity for wind reached 2,559 MW by the end of 2005.[56]

The rapid growth of China's wind-energy sector can be accounted for by the operation of the Renewable Energy Law and a number of complementary policy measures since 2005. Of particular importance are the mandatory grid access that obligates utility companies to purchase all of the wind energy produced and to connect wind farms to the grid;[57] feed-in tariffs for wind-power generation, introduced nationwide in 2009, which can apply to a farm's entire operational period (usually 20 years), with four different tariff categories (ranging from US$0.08/kWh to US$0.1/kWh), set substantially higher than through the bidding programme of

49 Ibid.

50 Qiang Wang, 'Effective Policies for Renewable Energy – Example of China's Wind Power – Lessons for China's Photovoltaic Power' (2010) 14 *Renewable and Sustainable Energy Reviews* 702, 706.

51 Ibid.

52 Zhang (n 36), 6643–6644.

53 Wang Zhongying et al., 'The Retrospective Review and Assessment of China's Renewable Energy Law Implementation' (2006, NDRC Energy Research Institute and Beijing Oriental Environmental Research Institute), 28.

54 'Ride the Wind Programme' (乘风计划) was designed to initiate an ambitious localization programme to gradually increase the local content to 80%. The concession policy programme launched later mandated that wind turbines be manufactured with 70% local content.

55 IRENA, '30 Years of Policies for Wind Energy Lessons from 12 Wind Energy Markets' (IRENA, 2013) <www.irena.org/documentdownloads/publications/gwec_china.pdf> accessed 12 August 2019, 50.

56 Ibid.

57 See Renewable Energy Law 2005.

the early 2000s;[58] and capacity targets established in the Medium- and Long-Term Development Plan for Renewable Energy.[59] These targets were based on installed generating capacity rather than the actual amount of electricity connected to the power grid. These supportive measures sent a strong signal of long-term stability to wind investors. The annual growth rate of China's wind-power generation capacity in 2006 and 2007 exceeded 100%.[60] By the end of 2007, a total of 158 wind-power plants had been built in 21 Chinese provinces, with Inner Mongolia having the largest capacity.[61] China surpassed the United States in cumulative installed capacity in 2010, and with 75.56 GW installed in 2012, it accounted for more than one-quarter of world capacity in that year.[62]

The localization requirements remained in place during this period, which encouraged international players to establish manufacturing facilities in China, either as independent businesses or in joint ventures with local manufacturing companies.[63] In 2012, the government selected seven domestic manufacturers to supply over 5 GW of wind turbines to 25 projects in the provinces of Inner Mongolia and Hebei.[64] Four Chinese companies thereafter rose to the world's top-ten wind-turbine manufacturers and began to expand into foreign markets as increasingly competitive global players.[65]

China's use of local-content requirements in the wind sector met opposition from the United States, which led to a World Trade Organization consultation requested by the latter against China.[66] In response, China removed those requirements.[67]

58 The pricing setting used the earlier system of wind-farm bidding prices as a point of reference.

59 The target was set initially to achieve 30 GW of grid-connected wind capacity by 2020. However, in 2010, China had eclipsed the 2020 target ahead of schedule. The current target has been amended to be 200 GW by 2020.

60 Ni Chunchun, 'China's Wind Power Generation Policy and Market Developments' (Institute of Energy Economics, Japan, 2008), 4.

61 Ibid.

62 Sufang Zhang et al., 'The Development Trajectories of Wind Power and Solar PV Power in China: A Comparison and Policy Recommendations' (2013) 26 *Renewable and Sustainable Energy Reviews* 322, 324.

63 International wind-turbine manufacturing firms such as Gamesa, General Electric, Nordex, Suzlon, and Vestas established manufacturing facilities in China. See Joanna Lewis, 'Building A National Wind Turbine Industry: Experiences from China, India and South Korea' (2011) 5(3/4) *International Journal of Technology and Globalization*, 281.

64 IRENA, '30 Years of Policies for Wind Energy: Lessons from 12 Wind Energy Markets' (IRENA, 2013) <www.irena.org/documentdownloads/publications/gwec_china.pdf> accessed 12 August 2019, 52.

65 The four Chinese companies are: Sinovel, Goldwin, United Power, and Dongfang Electric.

66 Doug Palmer, 'US Challenges China Wind Power Aid at WTO' (23 December 2010, Reuters) <www.reuters.com/article/us-usa-china-windpower/u-s-challenges-china-wind-power-aid-at-wto-idUSTRE6BL3EU20101222> accessed 12 August 2019.

67 Notice of NDRC on Abolishing the Requirement on the Rate of Localization of Equipment Procurement on Wind Power Projects (国家发改委下发国家发展改革委员会关于取消风电工程项目设备采购国产化率要求的通知), NDRC Order No. (2009)2991, 25 November 2009.

The fast growth of wind power exerted pressure on China's grid infrastructure and led to delays in connecting wind-generation capacity to the grid.[68] The large-scale introduction of power from intermittent wind resources challenges grid stability. Other factors, such as local consumption, price competitiveness, and the rigidity of electricity dispatching systems, also contributed to the reluctance of Chinese grid companies to connect wind power to the grid.[69] The reality was that grid companies, having market dominance and state ownership, could avoid the obligation to purchase and distribute wind power, with near impunity.[70] Despite the 2005 Renewable Energy Law's imposition of the full-purchase requirement on grid companies and the granting to the State Electricity Regulatory Commission of the discretion to impose penalties on noncompliance, there have been no reported cases of penalties.[71] As a result, the curtailment rate of wind power in China has remained high, causing huge losses in energy resources, incurring extra costs for new-energy power generation, and hindering further adjustments to on-grid tariffs for new renewable energy.[72]

Measures used by China's central authorities to address wind-energy curtailment have included a renewable-energy development and use target guidance system and a wind-power surveillance and early-warning mechanism.[73]

As a new addition to the wind-power policy toolkit, surveillance and early-warning results are to be used to provide guidance to provinces and regions on wind-power development and investment. The warning degrees are categorized, from high to low, into three levels: Red, Orange, and Green. If the early-warning result is Red, the National Energy Administration (NEA) will not release any development projects within the year of the result, and localities will suspend the approval of new wind-power projects. If the result is Orange, the NEA in principle will not release any development projects for that year. The Green result means normal operation. Another policy has been to control the construction scale in northern regions where curtailment is serious and to

68 Zhang and others (n 10), 363.

69 Chen Jinqiang, 'The Challenges and Promises of Greening China's Economy' (January 2017, Harvard Kennedy School, Belfer Center for Science and International Affairs) <www.belfercenter.org/publication/challenges-and-promises-greening-chinas-economy> accessed 12 August 2019.

70 Iacob Koch-Weser, 'China's Wind and Solar Sectors: Trends in Deployment, Manufacturing, and Energy Policy' (US – China Economic and Security Review Commission Staff Research Report, 9 March 2015) 18.

71 Sara Schuman and Alvin Lin, 'China's Renewable Energy Law and Its Impact on Renewable Power in China: Progress, Challenges and Recommendations for Improving Implementation' (2012) 51 *Energy Policy* 89, 95.

72 Energy Research Institute of Academy of Macroeconomic Research and China National Renewable Energy Centre, 'China Renewable Energy Outlook 2017' (10 November 2017) <http://boostre.cnrec.org.cn/wp-content/uploads/2017/10/CREO-2017-EN-20171113-1.pdf> accessed 12 August 2019, 99. Provinces and municipalities with a high rate of wind power curtailment are: Inner Mongolia, Heilongjiang, Jilin, Ningxia, Gansu, and Xinjiang.

73 See Notice of NEA on Establishing a Surveillance and Early Warning Mechanism for Promoting Sustainable and Healthy Development of the Wind Power Industry (国家能源局关于建立监测预警机制促进风电产业持续健康发展的通知), NEA Order No. (2016)196, 18 July 2016.

shift the focal areas of wind-power development to other regions.[74] Yet another policy has been to optimize dispatching and operation, improve peak-load regulation and new-energy absorption capacity, and promote full-power generation from clean-energy sources. Lastly, tariffs for wind-power projects have been limited, with the NEA deciding to cease the approval of new wind projects in regions with an existing wind curtailment rate higher than 20% and to establish a renewable-energy quota system that requires a certain percentage of electricity consumption to be purchased from non-hydro renewable energy sources, including wind.[75]

China's strategy to boost wind-energy development since 2005 has met with a high degree of success, as measured by installation and manufacturing capacities, which attests to the positive correlation between a supportive legal and policy framework and the growth of renewable energy. However, the weaknesses associated with the design and implementation of wind-energy policies cannot be downplayed in light of the high rate of wind-power curtailment in China.

Overview of development in law and policy in China's solar-energy sector

China has rich solar resources, with the radiation in more than two-thirds of the country above 5,000 MJ/m^2 per year.[76] It represents a great growth potential for the solar-energy sector. The strong solar PV development in China has been discussed extensively in the literature in several different fields, including policy, economics, and political economy.[77] The linkage between solar-policy measures and solar development in China warrants examination.

74 Ibid.

75 Qi Ye, Jiaqi Lu, and Mengye Zhu, 'Wind Curtailment in China and Lessons from the United States' (Brookings-Tsinghua Center for Public Policy, 2018) <www.brookings.edu/wp-content/uploads/2018/03/wind-curtailment-in-china-and-lessons-from-the-united-states.pdf> accessed 12 August 2019.

76 Philip Andrews-Speed and Sufang Zhang, *China as a Global Clean Energy Champion: Lifting the Veil* (Springer, 2019), 114.

77 See, e.g., Chen Gang, 'China's Solar PV Manufacturing and Subsidies from the Perspective of State Capitalism' (2015) 33(10) *Copenhagen Journal of Asian Studies* 90; Sufang Zhang, Philip Andrews-Speed, and Meiyun Ji, 'The Erratic Path of the Low-Carbon Transition in China: Evolution of Solar PV Industry' (2014) 67 *Energy Policy* 903; Sufang Zhang and Yongxiu He, 'Analysis of the Development and Policy of Solar PV Power in China' (2013) 21 *Renewable and Sustainable Energy Reviews* 393; Ping Huang et al., 'How China Became a Leader in Solar PV: An Innovation System Analysis' (2016) 64 *Renewable and Sustainable Energy Reviews* 777; Zhenyu Zhao, Shuang-ying Zhang, and Jian Zuo, 'A Critical Analysis of the Photovoltaic Power Industry in China – From Diamond Model to Gear Model' (2011) 15(9) *Renewable and Sustainable Energy Reviews* 4963; Zhong Shuying, Liu Chi, and Qin Liqiong, 'Solar Industry Development and Policy Support in China' (2011) 5 *Energy Procedia* 768; Liu Liqun et al., 'Solar Energy Development in China – A Review' (2010) 14(1) *Renewable and Sustainable Energy Reviews* 301; Hongyang Zou et al., 'Market Dynamics, Innovation, and Transition in China's Solar Photovoltaic (PV) Industry: A Critical Review' (2017) 69(C) *Renewable and Sustainable Energy Reviews* 199.

In the early stage of solar development in China, in the 1990s, the central government's priority was to provide electricity to off-grid rural areas.[78] Schemes such as the Brightness Program, the Township Electrification Program, and the Village Electrification Program are examples.[79] Such schemes remained small in scale and contributed little to the development of domestic manufacturing or innovation in the solar-energy sector. The cost of solar PV was high, and it restricted the market's expansion in China.

China's solar PV manufacturing sector experienced substantial growth and innovation beginning in the early 2000s, due to Europe's enactment of supportive measures, such as feed-in tariffs.[80] Export-led growth boosted the global market share of Chinese PV manufacturing firms from around 1% in 2001 to around 45% in 2010.[81] The global financial crisis in 2008 dented prospects for the industry, as did the ensuing phase-out of favourable solar-market policies in Europe. Following this, US and European governments initiated anti-dumping investigations on Chinese solar exports and imposed high tariffs to squeeze Chinese manufacturers' market share.[82]

To rescue China's solar-manufacturing firms in a time of crisis, the Chinese government launched solar-deployment measures to increase domestic demand on a large scale – an area that until that time had been virtually ignored. The accumulated experience in China's wind-power sector helped inform the formulation of its solar-energy policy.[83] In 2009, the Ministry of Finance developed two initiatives – the Solar Roofs Program and the Golden Sun Demonstration Project – aimed at absorbing manufacturing capacity by subsidizing the installation of solar PV nationwide.[84] The programmes paid developers a percentage of the money that they had invested in solar projects, regardless of the amount of electricity being generated.[85]

78 Zhang and others (n 62), 326.

79 See NDRC Energy Research Institute, 'Research Report on China's Solar PV Industry Development' (中国光伏产业发展研究报告) (August 2006) <http://zsdh.library.sh.cn:8080/FCKeditor/filemanager/upload/jsp/UserImages/1216361096343.pdf> accessed 12 August 2019.

80 Jos Sijm, 'The Performance of Feed-in Tariffs to Promote Renewable Electricity in European Countries' (November 2002, ECN Project on Renewable Electricity Trends in European Countries) <www.ecn.nl/docs/library/report/2002/c02083.pdf?origin=publication_deta> accessed 12 August 2019.

81 China Global Trade, 'China's Solar Industry and the U.S. Anti-dumping/Anti-subsidy Trade Case' (May 2012) <www.chinaglobaltrade.com/sites/ default/files/china-global-trade-solar-manufacturing_may2012_0.pdf> accessed 12 August 2019.

82 Ministry of Commerce People's Republic of China, 'Timeline: China-EU, China-US Tug of War on Solar Duties' (19 June 2013) <http://english.mofcom.gov.cn/article/zt_solar/column2/201307/20130700218603.shtml> accessed 12 August 2019.

83 Koch-Weser (n 70), 15.

84 For more information on the two programmes, see Notice of MOF, MOST, and NEA on Golden Sun Demonstration Project (财政部，科技部和国家能源局关于印发金太阳示范工程财政补助资金管理暂行办法的通知), MOF, MOST and NEA, Order No. (2009)397, 21 July 2009; Notice of MOF and MOHURD on Solar Roofs Program (财政部和住房和城乡建设部印发关于实施太阳能屋顶计划的通知), MOF and MOHURD Order No. (2009)34, 16 April 2009.

85 Ibid.

The government also initiated two rounds of solar PV concession programmes for large-scale PV, which not only promoted China's installed capacity of PV power but also contributed to driving down the on-grid price of PV power in the country.[86] In 2011, the NDRC introduced the first nationwide feed-in-tariff scheme for solar PV power. Its point of reference was the 2009 feed-in tariff in Jiangsu province, the area in China with the largest number of solar-manufacturing firms.[87] The national feed-in tariff was RMB1.15/kWh for projects completed by 31 December 2011 and RMB1.0/kWh for projects approved by July 2011 but not completed before end of that year.[88] China's State Grid took action to expand and upgrade its grids to provide better connections for solar power. It included a plan to allow distributed solar-power generators smaller than 6 MW to connect to its power lines.[89]

The problem of curtailment that occurred in China's wind-energy development emerged in the PV sector as well, causing a bottleneck threatening the sustained development of solar energy in China.[90] In 2017, 11.5% of total solar-power generation was unused – an increase of 4.3% on the previous year.[91] The reasons are as follows.[92] The unified benchmark price set for the solar feed-in tariff nationwide drove project developers to install solar power in western regions of the country that have abundant solar resources yet low energy demand.[93] The imbalance of solar-power supply and demand became acute.[94] With the rapid reduction in the cost of solar equipment and power generation, the subsidy rate, which was not adjusted accordingly, turned out to be overgenerous, leading to excessive investment in installation. On the grid-connection side, not only did China not solve the problem of insufficient transmission lines across provinces and regions but some local governments even erected administrative barriers to protect the local thermal industry and restrict the import of power generated in other provinces.[95] In other cases, efforts to expand long-distance power-transmission

86 Zhang, Andrews-Speed, and Ji (n 77), 909. The price of on-grid solar PV power was reduced from CNY 4 per kWh in 2008 to CNY 1 per kWh or less in 2010.

87 See NDRC, Notice on Perfection of Policy Regarding Feed-in Tariff for Solar PV Generation.

88 Ibid.

89 Xinhua, 'New Policies Support Ailing Solar Industry' (29 October 2012) <www.chinadaily.com.cn/business/2012-10/29/content_15853949.htm> accessed 12 August 2019.

90 Ningning Tang et al., 'Solar Energy Curtailment in China: Status Quo, Reasons and Solutions' (2018) 97 *Renewable and Sustainable Energy Reviews* 509, 514.

91 NEA, 2016 Northwest Renewable Energy Integration Report (2016年西北区域新能源并网ff行情况通报) (19 January 2017) <www.nea.gov.cn/2017-01/19/c_135996630.htm> accessed 22 January 2019.

92 Tang et al. (n 90), 515–518; Shiyu Liu et al., 'Curtailment of Renewable Energy in Northwest China and Market-based Solutions' (2018) 123 *Energy Policy* 494, 496–497.

93 Zeng Ming et al., 'Overall Review of Renewable Energy Tariff Policy in China: Evolution, Implementation, Problems and Countermeasures' (2013) 25 *Renewable and Sustainable Energy Reviews* 260, 269.

94 Tang et al. (n 90), 515.

95 Liu (n 92), 496–497.

infrastructure were blocked, out of concern that local power generators would be disadvantaged while producers in the western areas would benefit.[96]

Government bodies such as the State Council, NDRC, and NEA have responded to these problems. First, a number of policies have been put in place to support distributed PV power at an early development stage, which can alleviate the mismatch between the geographic distribution of solar-power resources and power consumption.[97] Second, outdated subsidies given to solar-power plants have been phased out.[98] For instance, the government cut the feed-in tariff rate for utility-scale solar projects by about 20%.[99] Another example is the implementation of a programme by the NEA to enhance technological progress, industrial upgrading, cost reduction, and grid parity in the solar sector.[100] Instead of focusing merely on the amount of installed capacity, that programme (called Top-Runner) gave preference to solar projects that are more technologically advanced and cost-competitive.[101] Third, the NEA halted approvals for new solar projects in regions with a high rate of solar-power curtailment in 2019.[102]

Conclusion

The development of renewable energy has become a worldwide trend to meet the demands of facilitating energy transition, enhancing industrial competitiveness, and tackling climate change. In acknowledgement of this, China's government promulgated the Renewable Energy Law in 2005 and amended it in 2009, thus creating the basic legal system for renewable-energy development in the country. Numerous supportive rules and policies were adopted at central and local levels in China to enhance and complement the implementation of Renewable Energy Law. Through statute and policy, China's transition to a low-carbon economy on an unprecedented scale has been promoted, particularly in terms of the installation

96 Jeffrey Ball, 'Grow Green China Inc. How China's Epic Push for Cleaner Energy Creates Economic Opportunity for the West' (Brookings, May 2019) <www.brookings.edu/research/grow-green-china-inc/> accessed 22 January 2020.

97 Zhang, Andrews-Speed, and Ji (n 77), 911.

98 Xinhua, 'Fading Subsidies, Rising Sustainability in China's Solar Energy Industry' (23 July 2019) <http://global.chinadaily.com.cn/a/201907/23/WS5d36c13ca310d830564008fb.html> accessed 22 January 2020.

99 Ibid.

100 Notice of NEA on Promoting the Top-Runner Program and Requirements of Implementing the Program in 2017 (国家能源局关于推进光伏发电领跑者计划实施和2017年领跑基地建设有关要求的通知), NEA Order No. (2017)54, 22 September 2017.

101 In 2016, there were eight main Top-Runner bases that adopted a public competitive bidding mode for grid connection. On average, each project saw a 0.2-yuan price reduction compared to the local benchmark on-grid tariff of solar PV electricity. The eight Top-Runner bases were located in five provinces: Hebei (500MW), Shanxi (1GW), Inner Mongolia (1.5GW), Anhui (1GW), and Shandong (1GW), with a total capacity of 5.5GW.

102 David Stanway, 'China Blocks New Solar in 3 NW Regions amid Overcapacity Fears', *Reuters* (15 February 2015) <www.reuters.com/article/us-china-solarpower/china-blocks-new-solar-in-3-nw-regions-amid-overcapacity-fears-idUSKCN1Q404G> accessed 22 January 2020.

and manufacturing capacity in the renewable-energy sector. Given the magnitude of its potential impact on climate change, China's renewable-energy development can be an important building block in achieving sustainable development on a global scale.

Although China has made remarkable progress in accelerating the deployment and development of renewable energy, its renewable-energy legal and policy framework is not yet comprehensive or fully developed, which leaves significant room for improvement. Recent years have witnessed the emergence of factors impeding the sustained development of renewable energy, especially in the wind- and solar-energy sectors, such as a lack of connectivity between renewable-energy power and the grid system; the unrestricted installation of renewable energy in regions with low demand; and a manufacturing capacity surplus. These issues underscore the deficiencies in China's renewable-energy laws and policies. Addressing these problems presents a great challenge, but also a great opportunity for China to realize a diverse, large-scale, and stable renewable-energy portfolio capable of meeting its rapidly increasing energy demands and mitigating climate change.

5 China's legal framework for emission trading and other market initiatives

Hao Zhang

Introduction

This chapter examines the legal framework for emission trading and other market initiatives that aim to reduce greenhouse gas (GHG) emissions in China. Market-based instruments such as emission-trading schemes (ETSs) are part of the mitigation tools that complement the existing command-and-control measures for decarbonization.[1] At present, the market initiatives that underpin China's mitigation strategies consist of emission trading and a renewable-energy quota system. China's legal framework for emission trading has been gradually developed through a strategy of developing the national scheme through local pilot programmes.[2] The pilots are designed and operated to test regulatory design and implementation, with the ultimate goal of facilitating the design of the national emission trading programme.[3] The renewable-energy quota system, initiated in 2018, is a relatively new market instrument that mandates priority access for renewable energy to the grid network through binding targets and trading renewable-energy certificates (RECs).[4] RECs are issued to renewable-energy generators for the amount of electricity they produce. With binding targets that apply to the provinces, provincial governments are responsible for achieving the allocated targets; within each province, liable entities are obliged to purchase RECs from the market as they are bound to prove that a certain amount of their electricity comes from renewable-energy sources.[5]

1 Anatole Boute and Hao Zhang, 'The Role of the Market and Traditional Regulation to Decarbonise China's Energy Supply' (2018) 30(2) *Journal of Environmental Law* 261.

2 Notice of the NDRC on Initiating Pilot Emissions Trading Programs (国家发展改革委办公厅关于开展碳排放权交易试点工作的通知) NDRC Order No. (2011)2601, 29 October 2011.

3 Ibid; see also Hao Zhang and Ping Xu, 'Designing Regulation for China's Emission-Trading Pilot Programs Through Trial and Error: An Effective Approach?' (2017) 7(2–3) *Climate Law* 125.

4 For the binding target, see Notice of NDRC and NEA on Establishing and Improving the Safeguard Mechanism for Renewable Energy Consumption (国家发展改革委 国家能源局关于建立健全可再生能源电力消纳保障机制的通知), NDRC Order No.(2019)807, 10 May 2019. For the RECs system, see the Measure for Renewable Electricity Quota and Assessment (Draft for Comment) (可再生能源电力配额及考核办法(征求意见稿)), National Energy Administration Order No. (2018)28, 23 March 2018.

5 Ibid.

Given the nascency of the two market-based instruments in China, this chapter discusses the two systems separately. It focuses on the legal obstacles that hinder their functioning and implementation and on the potential impact of the two instruments on each other as it relates to their effectiveness. The chapter thus aims to reveal the legal framework of the two key regulatory instruments in China and contribute to scholarly discussion regarding the role of the market-based instruments in China in the context of climate mitigation.

The chapter is structured as follows. The next section provides a contextual background on the legal and regulatory framework for emission trading and the renewable-energy quota system in China. The subsequent section examines the key regulatory aspects required to support the functioning of the two systems, with a focus on the challenges and barriers that prevent the two systems from achieving the desired regulatory goals.

Regulatory framework for emission trading and the renewable-energy quota system: context and development

China's pledge at the Copenhagen COP in 2009 included a 2020 target to reduce the intensity of carbon emissions by 40%–45% from 2005 levels.[6] Subsequently, the National Development and Reform Commission (NDRC),[7] China's central planning powerhouse for climate policy, took steps to regulate and reduce China's GHG emissions.[8] Despite the fact that that target was not binding under the UNFCCC process, it was made binding at the domestic level, as the NDRC tried to steer China's development path to be more friendly to climate and environment. To achieve the target by 2020, incremental targets were adopted on the basis of five-year intervals, aligned with China's five-year planning system.[9]

6 Letter from Director General of Department of Climate Change, National Development and Reform Commission, China, to Executive Secretary (UNFCCC Secretariat, Germany, 28 January 2010) <unfccc.int/files/meetings/application/pdf/chinacphaccord_app2.pdf> accessed 4 September 2019.

7 Due to the government restructuring in 2018, the Department of Climate Change was moved from the NDRC to the Ministry of Ecology and Environment. As observed by some commentators, this has led to delays in emission-trading legislation and the launch of the national ETS: 'China National ETS' (International Carbon Action Partnership, April 2019) <https://icapcarbonaction.com/en/?option=com_etsmap&task=export&format=pdf&layout=list&systems%5B%5D=55> accessed 4 September 2019; Paul A. Davies and R. Andrew Westgate, 'The Chinese Government Announced Measures to Assist With ETS Implementation' (Latham & Watkins LLP, July 2018) <www.globalelr.com/2018/07/the-chinese-government-announced-measures-to-assist-with-ets-implementation/> accessed 4 September 2019.

8 Work Plan to Control Greenhouse Gas Emissions during the 12th Five-Year Period ("十二五"控制温室气体排放工作方案), State Council Order No. (2011)41, 1 December 2011; Notice of the State Council on Issuing the Work Plan to Control Greenhouse Gas Emissions during the 13th Five-Year Period (国务院关于印发"十三五"控制温室气体排放工作方案的通知), State Council Order No. (2016)61, 27 October 2016.

9 Ibid.

During the period of the 11th Five-Year Plan (2006–2010),[10] targets to reduce energy consumption were proposed by Chinese provinces, and specific targets were then allocated to each province after adjustments made by the central government to ensure that a national target would be reached.[11] However, an assessment of target achievement at the end of that period suggests that the target was narrowly missed, raising questions about the ability of command-and-control measures to achieve the desired reduction targets in the long run.[12] Fundamentally, the target failure enabled the NDRC to transform China's climate mitigation strategies in such a way as to encourage innovation and flexibility and to provide greater incentives to induce participation by industry.[13] As observed by some commentators, an emission-trading scheme has the advantage of flexibility in accommodating economic development (e.g. through setting a relatively loose cap) and being easier to adapt to China's existing regulatory and administrative framework.[14]

The development of renewable energy is another important area for China's mitigation strategy. By leveraging its renewable-energy resources during its participation in the Kyoto Protocol's Clean Development Mechanism, China has benefited significantly from deploying renewable-energy generation.[15] With the generation assets now being the largest in the world, China is facing a serious issue of renewable-energy curtailment.[16] Studies suggest that the challenge remains whether the regulatory instrument to support the integration of renewable energy can create sufficient incentives to drive the transformative change in China's power system, because past efforts have so far failed to address the issue of incentives for consumption.[17] Given that China's power system is now

10 Written Reply on Allocated Target of Reducing Energy Consumption per Unit GDP at Provincial Level during the Eleventh Five-Year Plan Period (国务院关于"十一五"期间各地区单位生产总值能源消耗降低指标计划的批复) State Council Order No. (2006)94, 17 September 2006. The first target was defined as energy intensity, as opposed to carbon-emission intensity (also referred to as carbon intensity), the latter of which has been used since the 12th Five-Year Plan period.

11 Hao Zhang, 'China's Energy Conservation and Carbon Emissions Reduction System: Development and Status Quo of the Regulatory and Institutional Framework' (2012) 42(3) *Environmental Law Reporter* 10260.

12 Ibid.

13 Ibid.

14 Zhongxiang Zhang, 'Carbon Emissions Trading in China: The Evolution from Pilots to a Nationwide Scheme' (2015)15 *Climate Policy* 104; Da Zhang, Valerie J. Karplus, Cyril Cassisa, and Xiliang Zhang, 'Emissions Trading in China: Progress and Prospects' (2014) 75 *Energy Policy* 9.

15 John A. Sautter, 'The Clean Development Mechanism in China: Assessing the Tension between Development and Curbing Anthropogenic Climate Change' (2009) 27(1) *Virginia Environmental Law Journal* 91.

16 Sara Schuman and Alvin Lin, 'China's Renewable Energy Law and its Impact on Renewable Power in China: Progress, Challenges and Recommendations for Improving Implementation' (2012) 51 *Energy Policy* 89; Sufang Zhang, Phillip Andrews-Speed, and Sitao Li, 'To What Extent Will China's Ongoing Electricity Market Reforms Assist the Integration of Renewable Energy?' (2018) 114 *Energy Policy* 165.

17 Fredrich Kahrl, Jim Williams, Ding Jianhua, and Junfeng Hu, 'Challenges to China's Transition to a Low Carbon Electricity System' (2011) 39 *Energy Policy* 4032.

undergoing some structural changes, in particular a reform of the electricity sector since 2015 that aims to deregulate the sector to achieve higher efficiency and flexibility,[18] it is still questionable whether the newly established renewable electricity quota system can improve the power-system management in China (such as the inter- and cross-provincial trading of renewable electricity).[19] This issue is discussed in more detail below.

Emission trading

As a key strategy of China's climate mitigation policy, the local emission-trading pilot programmes have been in operation in four municipalities (Beijing, Tianjin, Shanghai, and Chongqing), two provinces (Hubei and Guangdong), and one 'independent planning' city (Shenzhen).[20] The seven pilot programmes have some commonalities, as all of them are based on the cap-and-trade model.[21] But they differ significantly in terms of their design and the choices made for each of the regulatory elements under the cap-and-trade model, as well as their starting dates (as shown in Table 5.1).[22]

Since the launch of China's first pilot programme in Shenzhen in June 2013, China's experience with emission trading has provided useful insights into the dynamics and interaction of national policy and local regulations.[23] National policy in this area is developed by defining the key parameters of the policy objectives, such as the national target;[24] it then authorizes the provincial/municipal Development and Reform Commissions (hereinafter referred to as local regulators) to begin experimenting with different ways to reach the target.[25] For local regulators selected for this task, implementation involves designing the laws and regulations for local emission-trading pilots and keeping these pilot programmes alive.[26]

18 Boute and Zhang (n 1).

19 Hao Zhang, 'Antinomic Policy Making under the Fragmented Authoritarianism: Regulating China's Electricity Sector through the Energy-Climate-Environment Dimension' (2019) 128 *Energy Policy* 162.

20 NDRC (n 2). Fujian province has opted to be part of the pilot programmes since December 2016 but this fact is generally is less known.

21 Zhang (n 14); Da Zhang et al., 'Emissions Trading in China: Progress and Prospects' (2014) 75 *Energy Policy* 9.

22 Maosheng Duan, Tao Pang, and Xiliang Zhang, 'Review of Carbon Emissions Trading Pilots in China' (2014) 25(3) *Energy and Environment* 527.

23 Frank Jotzo and Andreas Löschel, 'Emissions Trading in China: Emerging Experiences and International Lessons' (2014) 75 *Energy Policy* 3.

24 The national target is intensity based and is not an absolute cap on China's emissions. The national target is to be achieved by allocating the targets to provincial governments. For the local pilot programmes, the provincial target also does not function as a cap. The local ETS pilot programmes are important means to achieving the provincial target.

25 Zhang (n 11).

26 Shaozhou Qi, Banban Wang, and Jihong Jiang, 'Policy Design of the Hubei ETS Pilot in China' (2014) 75 *Energy Policy* 31; Jing Jing Jiang, Bin Ye, and Xiao Ming Ma, 'The Construction of Shenzhen's Carbon Emission Trading Scheme' (2014) 75 *Energy Policy* 17.

Table 5.1 Overview of the design features of the ETS pilot programmes in China

Location	*Scope of Coverage*			*Permit Allocation*		*Price Regulation*	*Linkage with Offsets*		*Market Openness to Individual and Institutional Investors*	*Start Date*
	Emission Cap (Mt CO_2 eq.)	*Number of Covered Entities*	*Threshold*	*Main Method*	*Baseline Year Range for initial allocation*		*Volume*	*Source and Sector*		
Beijing	46	947	10,000 tons CO_2	Grandfathering	2009–2011	Safety valve, auction permit, and buy-back permit	5%	CCERs, 50% from local	Individual and Institutional	28 Nov 2013
Shanghai	156	298	20,000 tons CO_2 for industry; 10,000 tons CO_2 for non-industry	Grandfathering	2009–2012	Allowance reserve; price fluctuation limit	5%	CCERs	Institutional only	26 Nov 2013
Tianjin	160–170	109	20,000 tons CO_2	Grandfathering	2009–	Safety valve	10%	CCERs	Individual and Institutional	26 Dec 2013
Chongqing	100.4	237	20,000 tons CO_2; 10,000 tons CO_2 for non-industry	Grandfathering	2008–2010	20% price fluctuation limit	8%	CCERs	Individual and Institutional	19 Jun 2014
Hubei	257	344	60,000 tons of standard coal	Grandfathering	2009–2011	Allowance reserve (8% of the cap)	10%	CCERs from Hubei	Individual and Institutional, incl. foreign	2 Apr 2014
Guangdong	422	296	20,000 tons CO_2 and 10,000 tons standard coal equivalent	Grandfathering	2010–2012	Yes, but not specified	10%	CCERs, 70% from local	Individual and Institutional	19 Dec 2013

Shenzhen	31.45 (excl. building)	824	3,000 tons CO_2 for industry; large office buildings and government buildings with areas more than 10,000 m^2	Grandfathering	2009–2011	Allowance reserve; reserve for new entrants and buy-back permit	10%	CCERs, including CCERs from western China	Individual and Institutional, incl. foreign	18 Jun 2013

Notes: CCERs = China Certified Emission Reductions. Apart from Hubei and Guangdong, the information on the cap for other ETS pilots is based on estimation and secondary sources. Source: Compiled by author on the basis of official documents and secondary sources (n 21; n 22; n 30; n 52; n 75).

According to the ETS development roadmap issued by the NDRC, the national ETS was originally planned to start operation in 2017.[27] In December 2017, the NDRC announced that China's national ETS would be launched with coverage limited to the power sector.[28] This departed from the original plan, which covered eight sectors.[29] According to some commentators, this was due to a data-quality issue and a government restructure that shuffled the managing authorities for climate change.[30] In 2018, the Ministry of Ecology and Environment (MEE) was approved to replace the Ministry of Environmental Protection.[31] Before the restructure, the NDRC, in cooperation with other ministries, was in charge of policy design and rulemaking for the national ETS, and provincial Development and Reform Commissions (DRCs) were in charge of ETS pilots.[32] After the restructure, responsibility for climate change, including for the development of a national ETS, shifted from the NDRC to the MEE.[33] The major practical implication of the government reshuffle was that the momentum for developing the national ETS, particularly progress on developing national legislation on the ETS, was interrupted, largely due to the loss of expertise in key stakeholders affected by the restructure, which also affected coordination between the Department of Climate Change (part of the NDRC) and the MEE on issues related to climate and environmental policy.[34]

An expectation has built up that China's adoption of an ETS will control its GHG emissions effectively.[35] However, given the size and complexity of China's

27 The NDRC adopted a framework regulation on ETS, titled Interim Measures for the Administration of Carbon Emissions Trading (碳排放权交易管理暂行办法), NDRC Order No. (2014)17, 10 December 2014. Prior to the government restructure in 2018, the former Department of Climate Change under the NDRC initiated an internal consultation for the drafted legislation on the national ETS, known as the Carbon Emissions Trading Management Regulations (Review Draft) (碳排放权交易管理条例(送审稿)), with the aim of having the legislation adopted by the State Council before the commencement of the national ETS in 2017.

28 Notice of the NDRC on Issuing the Plan for Constructing the National Emissions Trading Scheme (Electricity Generation Sector) (国家发展改革委关于印发《全国碳排放权交易市场建设方案（发电行业）》的通知), NDRC Order No. (2017)2191, 18 December 2017.

29 Regulations on Managing the Emissions Trading Scheme (Review Draft) (碳排放权交易管理条例(送审稿)), NDRC No. (2016), 2016.

30 Chen Liu and Yao Song, 'Progress of China's ETS and Some Recommendations' (中国碳排放权交易市场建设现状与建议) (2019) 4 *International Petroleum Economics* (国际石油经济) 47; *Progress of China's Carbon Market 2017* (2017中国碳市场进展) (Energy Research Institute of NDRC and Environmental Defense Fund Report, 2017) <www.cet.net.cn/uploads/soft/201805/1_31184021.pdf> accessed 17 September 2019.

31 Notice of the State Council on the Establishment of Institutions of the Central Government (国务院关于机构设置的通知), State Council Order No. (2018)6, 22 March 2018.

32 Notice of the NDRC on Initiating Pilot Emissions Trading Programs (n 2).

33 Liu and Song (n 30).

34 The Stumbling Blocks in China's National ETS Need a Break-Through (全国碳市场建设诸多"拦路虎"待破) (Xinhua News, 2018) <https://baijiahao.baidu.com/s?id=1620423038236671908&wfr=spider&for=pc> accessed 17 September 2019.

35 Department of Climate Change of the MEE, 'Progress of Constructing China's ETS and Considerations for Next Steps' (中国碳排放交易市场建设进展及下一步工作考虑) (2018) 46 *Environmental Protection* (环境保护) 14.

national ETS, the central government in general holds a cautious attitude towards it.[36] In a broader context, global and national socio-political and economic forces have affected the speed and direction of China's ETS. The most significant of these include the ongoing and heightened US–China trade tensions and the threat of recession in China and more broadly in the world economy.[37] Domestically, China has been suffering from an economic slowdown since around 2012 that has affected market confidence.[38] This, combined with the realization of the complexity involved in successfully launching an ETS, is likely to have slowed down its development.[39]

Renewable-energy quota system

In March 2018, the NEA issued the Measure for Renewable Electricity Quota and Assessment, a regulatory instrument that assigns quotas for hydro and non-hydro renewable electricity consumption to each province in China.[40] The renewable-energy quota system mandates priority access of renewable energy to the grid network through two critical components that are integral to the market-based instrument, namely binding targets and the trading of RECs.[41] In essence, RECs are issued to renewable-energy generators for each MWh of electricity they generate, and there are separate obligations that apply to the provincial governments under targets for hydropower and other types of renewable energy, such as wind, solar, and biomass (see Tables 5.2 and 5.3).[42]

The system requires selected participants in China's electricity market, including grid companies, electricity retail companies, and large end users that take part in direct trading of power, to purchase a certain amount of RECs in order to demonstrate that a percentage of their electricity comes from renewable sources.[43] As stated in the Measure's objective, the renewable-energy quota system aims to "tackle the institutional constraints that are rooted in the pre-reform era".[44] The institutional constraints referred to consist of the decentralized energy governance

36 Liu and Song (n 30).

37 Mandy Meng Fang, 'A Crisis or An Opportunity? The Trade War between the US and China in the Solar PV Sector' (2020) *Journal of World Trade* forthcoming.

38 Ye Qi and Xiliang Zhang, *Annual Review of Low-Carbon Development in China 2018* (中国低碳发展报告2018) (Social Science Academic Publishing, 2018).

39 At the time of writing this chapter, the MEE was collecting information from the liable entities to prepare for the operation of the national ETS. See Notice on Compiling the List of the Liable Entities in the Power Sector for the National ETS and Surrendering Relevant Materials (关于做好全国碳排放权交易市场发电行业重点排放单位名单和相关材料报送工作的通知), MEE Order No. (2019)528, 27 May 2019.

40 Art. 2, Measure for Renewable Electricity Quota and Assessment (n 4).

41 Ibid.

42 Annex 2: Explanation of the Measure for Renewable Electricity Quota and Assessment, NEA (n 4). For the detailed information about the targets, see Notice of NDRC and NEA on Establishing and Improving the Safeguard Mechanism for Renewable Energy Consumption (n 4).

43 Measure for Renewable Electricity Quota and Assessment (n 4), art. 7.

44 Notice of NDRC and NEA on Establishing and Improving the Safeguard Mechanism for Renewable Energy Consumption (n 4), annex 2.

Table 5.2 Targets for hydropower renewable-energy consumption by province (autonomous region and municipality) in the overall electricity consumption

Provinces (Autonomous Regions and Municipalities)	*Minimum Target for 2018*	*Incentive Target for 2018*	*Minimum Target for 2019*	*Incentive Target for 2019*	*Minimum Target for 2020*	*Incentive Target for 2020*
Beijing	11.0%	12.1%	13.5%	14.9%	15.0%	16.5%
Tianjin	11.0%	12.1%	13.5%	14.9%	15.0%	16.5%
Hebei	11.0%	12.1%	13.5%	14.9%	15.0%	16.5%
Shanxi	15.0%	16.5%	15.5%	17.1%	16.5%	18.1%
Inner Mongolia	18.5%	20.4%	18.5%	20.4%	18.5%	20.4%
Liaoning	12.0%	13.2%	12.0%	13.2%	12.5%	13.7%
Jilin	20.0%	22.0%	21.5%	23.7%	22.0%	24.2%
Heilongjiang	19.5%	21.5%	21.5%	23.7%	26.0%	28.6%
Shanghai	31.5%	34.9%	32.0%	35.2%	33.0%	36.5%
Jiangsu	12.5%	13.7%	13.5%	14.9%	14.0%	15.5%
Zhejiang	18.0%	19.8%	17.5%	19.3%	18.5%	20.4%
Anhui	13.0%	14.3%	13.5%	14.9%	14.5%	15.9%
Fujian	17.0%	18.7%	18.5%	20.4%	19.5%	21.4%
Jiangxi	23.0%	25.1%	25.5%	28.1%	29.0%	32.1%
Shandong	9.5%	10.4%	10.0%	11.0%	10.0%	11.0%
Henan	13.5%	14.9%	13.5%	14.9%	16.0%	17.6%
Hubei	39.0%	43.0%	37.5%	41.3%	40.0%	44.0%
Hunan	46.0%	50.5%	47.0%	51.7%	49.0%	53.9%
Guangdong	31.0%	34.2%	28.5%	31.4%	29.5%	32.5%
Guangxi	51.0%	56.2%	45.5%	50.1%	50.0%	55.0%
Hainan	11.0%	12.1%	11.0%	12.1%	11.5%	12.6%
Chongqing	47.5%	52.1%	42.5%	46.8%	45.0%	49.5%
Sichuan	80.0%	88.0%	80.0%	88.0%	80.0%	88.0%
Guizhou	33.5%	36.9%	31.5%	34.7%	31.5%	34.7%
Yunnan	80.0%	88.0%	80.0%	88.0%	80.0%	88.0%
Tibet	–	–	–	–	–	–
Shaanxi	17.5%	19.2%	18.5%	20.4%	21.5%	23.7%
Gansu	44.0%	48.4%	44.0%	48.4%	47.0%	51.1%
Qinghai	70.0%	77.0%	69.5%	76.5%	70.0%	77.0%
Ningxia	20.0%	22.2%	20.0%	22.0%	22.0%	24.2%
Xinjiang	21.0%	23.1%	21.0%	23.1%	22.5%	24.5%

Source: Notice of NDRC and NEA on Establishing and Improving the Safeguard Mechanism for Renewable Energy Consumption, Annex 2 (n 4).

framework in which Chinese provincial governments and state-owned enterprises are endowed with considerable power and decision-making authority to shape the outcome of energy services and, in this case, renewable-energy consumption by end consumers.[45]

REC trading aims to resolve the issue of incentives by focusing on the demand side, which is a fundamental difference from the existing supporting systems

45 Zhang (n 19).

Table 5.3: Targets for non-hydropower renewable-energy consumption by province (autonomous region and municipality) in the overall electricity consumption

Province (Autonomous Region and Municipality)	*Minimum Target for 2018*	*Incentive Target for 2018*	*Minimum Target for 2019*	*Incentive Target for 2019*	*Minimum Target for 2020*	*Incentive Target for 2020*
Beijing	10.5%	11.6%	13.5%	14.9%	15.0%	16.5%
Tianjin	10.5%	11.6%	13.5%	14.9%	15.0%	16.5%
Hebei	10.5%	11.6%	13.5%	14.9%	15.0%	16.5%
Shanxi	12.5%	13.8%	13.5%	14.9%	14.5%	16.0%
Inner Mongolia	18.0%	19.8%	18.0%	19.8%	18.0%	19.8%
Liaoning	10.0%	11.0%	10.0%	11.0%	10.5%	11.6%
Jilin	15.0%	16.5%	15.5%	17.1%	16.5%	18.2%
Heilongjiang	15.0%	16.5%	17.5%	19.3%	20.5%	22.6%
Shanghai	2.5%	2.8%	3.0%	3.3%	3.0%	3.3%
Jiangsu	5.5%	6.1%	6.5%	7.2%	7.5%	8.3%
Zhejiang	5.0%	5.5%	6.5%	7.2%	7.5%	8.3%
Anhui	9.5%	10.5%	10.5%	11.6%	11.5%	12.7%
Fujian	4.5%	5.0%	5.0%	5.5%	6.0%	6.6%
Jiangxi	6.5%	7.2%	7.0%	7.7%	8.0%	8.8%
Shandong	9.0%	9.9%	10.0%	11.0%	10.0%	11.0%
Henan	9.0%	9.9%	9.5%	10.5%	10.5%	11.6%
Hubei	7.5%	8.3%	9.0%	9.9%	10.0%	11.0%
Hunan	9.0%	9.9%	11.5%	12.7%	13.0%	14.3%
Guangdong	3.5%	3.9%	3.5%	3.9%	4.0%	4.4%
Guangxi	4.0%	4.4%	4.5%	5.0%	5.0%	5.5%
Hainan	4.5%	5.0%	5.0%	5.5%	5.0%	5.5%
Chongqing	2.0%	2.2%	2.5%	2.8%	2.5%	2.8%
Sichuan	3.5%	3.9%	3.5%	3.9%	3.5%	3.9%
Guizhou	4.5%	5.0%	5.0%	5.5%	5.0%	5.5%
Yunnan	11.5%	12.7%	11.5%	12.7%	11.5%	12.7%
Tibet	–	–	–	–	–	–
Shaanxi	9.0%	9.9%	10.5%	11.6%	12.0%	13.2%
Gansu	14.5%	16.0%	17.0%	18.7%	19.0%	20.9%
Qinghai	19.0%	20.9%	23.0%	25.3%	25.0%	27.5%
Ningxia	18.0%	19.8%	18.0%	19.8%	20.0%	22.0%
Xinjiang	11.5%	12.7%	12.0%	13.2%	13.0%	14.3%

Source: Notice of NDRC and NEA on Establishing and Improving the Safeguard Mechanism for Renewable Energy Consumption, Annex 3 (n 4).

in China, which underpin the development of renewable energy on the supply side.[46] Realizing that the existing regulatory mechanisms have so far failed to address the issue of incentives, the NDRC and the NEA highlighted, in the Implementation Measures to Resolve the Curtailment of Hydropower, Wind, and Solar Energy, that the renewable electricity quota system will promote a shared

46 The supporting systems established under China's renewable energy law framework.

responsibility among the key stakeholders in the renewable-energy market.[47] Along with the minimum targets for provincial renewable-energy consumption that are established by the central government, the renewable electricity quota system promotes target achievement at the provincial level by trading RECs and subsequently showing that there are sufficient RECs to prove the achievement of the target.[48] In theory, the REC trading market will have the virtue of breaking down provincial boundaries to promote trading in renewable energy across provinces by imposing binding targets of renewable-energy quotas for provincial authorities to increase the consumption of electricity from renewable energy.[49] Nonetheless, as a crucial component of the REC trading market, the level of renewable-energy targets across provinces need to be fixed at a reasonable level. These targets need to be high enough to provide incentives for provinces with high electricity demand to absorb renewable electricity from heavily curtailed provinces, such as Gansu and Xinjiang. That having being said, it must also be recognized that setting reasonable obligations for each province requires a thorough understanding of the status quo of the provincial energy structure and the long-term target of renewable-energy consumption. Similar to the design and implementation of environmental targets in China, the process of establishing reasonable obligations and subsequently monitoring them is difficult – as further discussed next.

Barriers and obstacles to the functioning of market-based instruments

The ETS and the REC trading market share commonalities but also differences in the challenges and barriers to their functionality. First, as "markets" being created by regulators, these two systems are challenged by the supply and demand balance, which is often a direct result of the regulatory choices made by regulators. For the ETS, the cap determines the total amount of supply of allowances in the market, and the allocation process dictates how many allowances each liable entity is entitled to and therefore their demand for allowances during each compliance period. Because the national ETS is still under construction, the discussion in this subsection will focus on the experiences from selected ETS pilots. For the REC trading market, the supply of RECs is determined by how much electricity from renewables has been integrated into the grid network, whereas the demand is based on how stringent the targets are for absorbing electricity from renewable sources. Second, as an "artificial market" established by regulation, the issue of trading is a common obstacle to both systems because there are different levels

47 Implementation Measures to Resolve the Curtailment of Hydro, Wind and Solar Energy (解决弃水弃风弃光问题实施方案), NDRC and NEA Order No. (2017)1942, 8 November 2017.

48 Ibid.

49 Measure for Renewable Electricity Quota and Assessment (n 4), art. 23.

of regulation that govern the aspect of trading for the ETS and electricity market. In China, the lack of coordination between these regulatory regimes means that the trading activities are often confined to a certain limit, thus affecting the role of price signals that emanate from these two mechanisms, as well as their viability in practice. Because these two aspects are fundamental to the functioning of both mechanisms, the following discussion will focus on them and further elaborate the specific challenges that could undermine the regulation-making process or its implementation.

Emission trading

Cap setting and allowance allocation by the ETS pilots

Cap setting and allowance allocation have been approached differently by the ETS pilots. There are two approaches – absolute (and adjustable) cap, and intensity-based cap. These are represented in the ETS design of Hubei and Shenzhen, respectively. In the experience of Hubei and Shenzhen, the main challenge is that regulators tend to balance the contradictory regulatory goals (creating space for future emission growth while controlling emissions) by using a complex regulatory design in cap setting and allowance allocation. In practice, such complex arrangements lead to the over-allocation of allowances and regulatory uncertainty for liable entities.

In Hubei, the cap is set by using a combination of historical and predictive methods, taking into account the provincial target of reducing emission intensity in projected GDP growth.[50] To reflect these considerations, the total cap in Hubei has three components: existing installations, new entrants, and a government reserve of allowances.[51] The historical emissions of covered entities in the baseline year are the basis on which the initial allocation to installations is decided, except in the case of cement, heating, and the power sector, where the benchmark method is used.[52] To ensure that the total number of allowances initially allocated to the covered entities is below the cap, Hubei adopts a cap-adjustment factor to reduce

50 Implementation Measures for Hubei's Emission Trading Pilot Program (湖北省碳排放权交易试点工作实施方案), Hubei People's Government Order (2013)9, 18 February 2013, part 3(2)(1).

51 See, e.g., Allocation Plan for the Allowances under Hubei's Emission Trading Pilot Program (湖北省碳排放权配额分配方案), Hubei DRC, 22 May 2014, part 3(2). The allocation plans in 2015 and 2016 maintained the same composition of the cap.

52 See, e.g., Appendix 1, Allocation Plan for the Allowances under Hubei's Emission Trading Pilot Program in 2015 (湖北省2015年碳排放权配额分配方案), Notice of the Hubei DRC on Issuing the Allocation Plan for the Allowances under Hubei's Emission Trading Pilot Program in 2015 (省发展改革委关于印发《湖北省2015年碳排放权配额分配方案》的通知), Hubei DRC Order No. (2015)708, 19 November 2015, part 4.

the total.[53] The predicted growth in emissions is used to set Hubei's limit of emission growth for new entrants and production expansion for existing facilities. For the purposes of price discovery, as well as price regulation of allowances, the government reserve of allowances is 8% of the overall cap.[54]

To control the risk of the over- or under-supply of allowances, Hubei has adopted an ex-post adjustment mechanism. This mechanism is implemented through the process of allowance allocation. The rule is that Hubei DRC has the authority to supplement or take back allowances where the verified emissions of an entity differ greatly from the number of allowances that were initially allocated.[55] Although allowances are generally allocated in full through grandfathering, a hybrid method of grandfathering and benchmarking applies to the cement producers, heating suppliers, and power generators.[56] In so doing, Hubei's pilot programme aims to achieve a slightly tighter allocation in the initial round. From the regulator's point of view, the risk of over-allocation poses greater regulatory challenges than does the shortage of allowances in the market, given the challenge of overcapacity in the aforementioned sectors.[57]

Shenzhen has taken the innovative approach of setting up an intensity-based cap on emissions.[58] Given the dynamic relationship between economic growth, structural transition, and emission control, the Shenzhen DRC has opted for an intensity-based cap so as to address the uncertainty in output levels from Shenzhen's rapid

53 In the first compliance period, the cap adjustment factor was 0.9192, and it was determined by the total emissions of the liable entities in 2010, multiplied by 97% then divided by the sum of the average emissions by the liable entities in the baseline years. In the subsequent compliance periods, the cap adjustment factor is determined by the following formula: the factor = 1 - (the amount of remaining allowances held by the liable entities from the previous year / the total number of allowances of the current year). The adjustment factors in 2015 and 2016 were 0.9883 and 0.9856, respectively. See, e.g., Appendix 1, Allocation Plan for the Allowances under Hubei's Emissions Trading Pilot Program in 2016 (湖北省2016年碳排放权配额分配方案), Notice of the Hubei DRC on Issuing the Allocation Plan for the Allowances under Hubei's Emissions Trading Pilot Program in 2016 (省发展改革委关于印发《湖北省2016年碳排放权配额分配方案》的通知), Hubei DRC Order No. (2016)791, 30 December 2016, part 4 (1)(3). Also, banking is not allowed unless the allowances have been transacted, meaning that if the remaining allowances held by the liable entities are obtained through trading, they can be banked for use in the following year. Otherwise, the allowances are cancelled at the end of the compliance year. This unique design of banking in Hubei's pilot programme is to encourage the trading of allowances, and thus liquidity in the market.

54 See, e.g., Allocation Plan for the Allowances under Hubei's Emission Trading Pilot Program (n 51).

55 Ibid. Each year before the end of the compliance period, the Hubei DRC will supplement or confiscate allowances if the difference between verified emissions and allocated allowances surpasses 20% or 200,000 tons. The amount to be supplemented or confiscated is the number of allowances that are beyond the threshold of 20% or 200,000 tons.

56 Allocation Plan for the Allowances under Hubei's Emission Trading Pilot Program in 2016 (n 53), part 4(3).

57 Qi, Wang, and Jiang (n 26).

58 Interim Measures for the Administration of Carbon Emission Trading in Shenzhen (深圳市碳排放权交易管理暂行办法), Shenzhen People's Government Order No. (2014)262, 19 March 2014, art. 10.

growth and accelerated structural adjustment.[59] Shenzhen's approach is in line with the national GHG-intensity target, which is broken down and assigned to provinces and municipalities every five years. The intensity-reduction target for Shenzhen is referenced to the target assigned to Guangdong province. In the first phase of Shenzhen's ETS pilot programme (2013–2015), the target, assigned to Shenzhen by Guangdong's provincial government, was to reduce emission intensity by 21% during the 11th Five-Year Plan period (2011–2015).[60] To achieve that goal, Shenzhen's pilot programme set intensity-reduction targets of, respectively, 2%, 2%, and 25% to the power, water supply, and manufacturing sectors.[61] For the power and water supply sectors, intensity is determined by the amount of CO_2 emitted per unit of electricity or water supplied.[62] The intensity-based cap for power and water suppliers is based on intensity benchmarks, combined with the projected level of output.[63] The manufacturing sector is subject to a sector cap established by intensity benchmarks and projected sector production.[64] In practice, determining intensity benchmarks through standardized or uniform products in the manufacturing sector poses great challenges. To alleviate this problem, the amount of CO_2 emitted per unit of industrial added value is used to establish intensity benchmarks, and it is combined with the projected sector output so as to determine the sector cap.[65]

The concept of intensity also underpins the initial round of allocating allowances, and the Shenzhen DRC applies another innovative method in this process. Free allowances are allocated to liable entities in the power and water supply sector through benchmarking,[66] and the allocation to covered entities in the manufacturing sector happens through competitive game theory.[67] The latter shifts the focus of allocation from a government-led approach to a game among manufacturers. With the sector cap given in the initial round of allocation, liable entities compete with each other in finite repeated rounds for free allowances.[68] Each year

59 China Emissions Exchange (深圳排放权交易所), Frequently Asked Questions under Shenzhen's Emissions Trading Program (深圳碳交易市场问答集), July 2013 <www.cerx.cn/u/cms/cn/201505/21162157si42.pdf> accessed 10 October 2019.

60 Ibid.

61 Jiang, Ye, and Ma (n 26).

62 See, for example, Notice of the Shenzhen DRC on Implementing the Carbon Emissions Trading Program in 2016 (深圳市发展和改革委员会关于开展2016年度碳排放权交易工作的通知), Shenzhen DRC, 18 September 2016, part 2.

63 Ibid.

64 Ibid.

65 Interim Measures for the Administration of Carbon Emission Trading in Shenzhen (n 58), art. 19, para. 2.

66 Ibid., art. 17, para. 1.

67 Ibid., art. 17, para. 2.

68 Ibid. In each round of the game, outcomes are notified to successful bidders, and the decisions are based on the competing firm's historical emissions, level of emissions in the given sector, commitment to reduction, and comparative advantage with other firms. A firm satisfied with the allocation result may choose to accept and leave the game. Entities that are dissatisfied continue to compete in the game in the next round. Since the cap is given, allowances accepted by entities will be deducted from the cap, and the number of allowances will reduce with the progressive rounds of the game. In the final round, firms that are still in the game can receive allowances only from the remaining balance.

before the end of the compliance period, Shenzhen DRC has the statutory authority to supplement or cancel allowances that were handed out for free during the initial round of allocation, taking into account the intensity target applicable to the sector and the realized level of output by the covered firms.[69] To balance demand and supply, the general principle used is that the number of supplementary allowances must be lower than the number of allowances that are subject to cancellation.[70]

Trading allowances

The trading of allowances in the pilot programmes exemplifies the barriers that result from the lack of central guidance on ETS operation. The barriers are twofold. On one level, due to the fact that the State Council has implemented strict regulations governing trading exchanges,[71] the carbon-emission exchanges, where trading of allowances takes place in China, are prohibited from operating like stock exchanges, thus affecting their operation, particularly on clearance of trading and transparency of allowance price.[72] To address this issue, different exchanges have promulgated different measures that facilitate trading and oversee trading activities.[73] However, the different measures, divergent practices, and lack of transparency in the ETS pilots in general make it difficult to assess their performance or discern lessons that are useful to the development of the national ETS.

The trading of pilot ETS allowances has been confined to the secondary market, and there is no market for derivatives.[74] Compliance buyers are particularly concerned that they cannot express their expectations of allowance prices in the future.[75] This restriction has posed significant challenges to their strategic investments

69 Ibid., art. 19, para. 1.

70 Ibid., art. 19, para. 3.

71 Implementation Opinions of the State Council on Rectification of Various Types of Trading Platforms (国务院办公厅关于清理整顿各类交易场所的实施意见), State Council Order No. (2012)37, 12 July 2012.

72 Ibid.

73 See, e.g., Interim Rules of Trading Carbon Emission Allowances at Beijing Environmental Exchange (北京环境交易所碳排放权交易规则(试行)) <http://images.bjets.com.cn/www/201501/20150113171115897.pdf> accessed 19 September 2019. Other pilot programmes have also adopted similar rules to oversee trading activities.

74 Because of the strict financial regulation on derivatives, the pilot programmes are not allowed to develop financial products related to emission allowances. See Interim Measures on Managing the Trading of Financial Derivatives Developed by Financial Institutions (金融机构衍生产品交易业务管理暂行办法), China Banking Regulatory Commission (adopted on 28 December 2006 and amended on 27 January 2011). To date, the pilot programmes in Shanghai and Hubei are given permission to trade in futures only. See, e.g., Shanghai Clearing House, Introduction to the Emission-Allowance Futures in Shanghai (上海碳配额远期产品介绍), 15 December 2016 <www.shclearing.com/sheaf/ywjs_sheaf/cpjs/201612/t20161215_211339.html> accessed 19 September 2019.

75 'Investigation Report of China Carbon Market 2016' (中国碳市场调查报告2016) (Environomist, 20 February 2016) <http://183.91.33.14/cache/www.environomist.com> accessed 20 September 2019, at 56–58.

relating to decarbonization. Consequently, liable entities tend to hold their allowances and avoid trading them unless they are confident that they hold what they need for compliance in the relevant compliance period.[76] Addressing these barriers requires the coordination between central government agencies to eliminate the restrictions on carbon-emission exchange and to promulgate new rules to allow for, and properly regulate, the trading of allowance derivatives.

Renewable-energy quota system

The supply and demand challenge

The regulatory failure of past experiments to integrate more electricity from renewable sources into the grid network suggests that a more practical question facing the policymakers in China is whether the regulatory mechanisms in use can create enough incentives to transform provincial electricity operation into a regional one, thus reducing the barriers emanating from provincial interest and local protectionism.[77]

In theory, the REC trading market, by imposing binding targets on renewable-energy quotas across provinces to increase consumption of electricity from renewable energy, will have the virtue of transcending provincial boundaries by encouraging the inter-provincial trading of renewable energy.[78] However, as mentioned earlier, for this to happen, the level of renewable-energy quotas for each province needs to be sufficiently high to provide incentives for provinces with high electricity demand to absorb renewable electricity from heavily curtailed provinces, such as Gansu and Xinjiang. Setting up reasonable obligations for each province requires the long-term planning of targets for renewable-energy consumption in China, and due to the general lack of appropriate planning in China's electricity sector, the process of establishing reasonable obligations (targets) has proven to be difficult.[79] Currently, the targets for 2018 (Tables 5.2 and 5.3) are too modest to drive a substantial increase of renewable-energy consumption by the end-consumers or mobilize further investments into the renewable-energy industry, because these targets are considered to be easily achievable.[80] There is a lack of a review mechanism in the REC trading rules for setting up future targets and for using the REC trading market to accelerate consumption and investment

76 Ibid.

77 Kahrl, Williams, Jianhua, and Hu (n 17); Schuman and Lin (n 16); Hao Zhang, 'Prioritizing Access of Renewable Energy to the Grid in China: Regulatory Mechanisms and Challenges for Implementation' (2019) 3 *Chinese Journal of Environmental Law* 167.

78 Anders Hove and Daniel Wetzel, 'China is Planning Provincial Quotas for Clean Energy' (China Dialogue, 2018) <www.chinadialogue.net/blog/10574-China-is-planning-provincial-quotas-for-clean-energy-/en> accessed 20 September 2019.

79 Genia Kostka, 'Command without Control: The Case of China's Environmental Target System' (2016) 10 *Regulation & Governance* 58.

80 Hove and Wetzel (n 78).

in renewable energy. In the long term, the REC trading market will benefit from a linear target that is increasing over time to provide certainty to renewable-energy investors and to avoid lock-in of coal-fired generation.

Trading RECs

Under the renewable-energy quota system, to reiterate, the primary obligations are that provinces are assigned a target for renewable-energy consumption and, within each province, grid companies, retail companies, and industrial end-users participating in direct trading of electricity are obliged to purchase a certain amount of RECs to prove that they have fulfilled their obligations.[81] Given that in 2018 China's installed capacity for renewable-energy generation accounted for around 38.3% of the total generation capacity,[82] achieving the targets of renewable-energy consumption of 15% by 2020 and 20% by 2030 will not require substantial new investment to be made in the coming years.

However, due to the structure of China's electricity system and also the existing regulatory arrangements governing the operation of the grid network, the challenge for REC trading is that the market is confined to the primary market, because the RECs depend on the flow of renewable electricity. Although there have been some efforts to reform the electricity sector in China, it is still heavily regulated.[83] Among the stakeholders (generators, grid operators, and end users), the grid operators still have exclusive control over the transmission and distribution network, particularly the dispatch.[84] In essence, grid companies are in a dominant position to decide the extent to which renewable energy can access the grid network for dispatch.[85] Given that the RECs are attached to the level of penetration of renewable energy in the grid network, achieving the target of renewable-energy consumption will depend largely on the dispatching of renewable energy by the grid operators. In light of the fact that the past reform efforts have so far failed to steer the dispatch to being more flexible and efficient, the rapid growth of renewable energy on the consumption side is beyond reach.[86]

The reality is that RECs that are dependent on the flow of renewable electricity will largely reduce their tradable value and also jeopardize the viability of the market itself. For the moment, the targets that have been assigned to provinces go

81 Measure for Renewable Electricity Quota and Assessment (n 4).

82 'The NEA Reported on the Operation and Integration of Renewable Energy in 2018' (国家能源局发布2018年可再生能源并网ff行情况等) (NEA, 28 January 2019) <www.gov.cn/xinwen/2019-01/28/content_5361939.htm#1> accessed 20 September 2019.

83 Kahrl, Williams, Jianhua, and Hu (n 17).

84 For more detailed information regarding electricity dispatch and its development in China, see Mun Ho, Zhongmin Wang, and Zichao Yu, 'China's Power Generation Dispatch' (Resources for the Future 2017), in Chinese <www.rff.org/files/document/file/RFF-Rpt-ChinaElectricity.pdf>; Fredrich Kahrl, James H. Williams, and Junfeng Hu, 'The Political Economy of Electricity Dispatch Reform in China' (2013) 53 *Energy Policy* 361.

85 Zhang (n 77).

86 Ibid.

up to 2020 (Tables 5.2 and 5.3).[87] As mentioned earlier, apart from the fact that the targets are considered modest,[88] there is a lack of long-term planning for them. The benefit of having more-stringent targets and a linear increase in future targets is that the integration of renewable energy into the grid, and thus the value of RECs, will improve. In addition, the nature of renewable-energy production (such as wind and solar) is that they fluctuate, and so will the number of RECs in the market. The yearly assessment will lead to the unintended consequence that trading of RECs (in the primary market) will spike before the year's end, and the trading volume, on average, will be small. A market without active trading of RECs will be unable to help the regulators understand the supply and demand of RECs (whether too scarce or too abundant), therefore reducing their capacity to adjust the quota and adversely impacting on the effectiveness of the target system overall.

Conclusion

Adoption of market-based instruments in China is, in general, a positive development that can help address several problems that are related to China's decarbonization efforts. These efforts include more integration of renewable energy into the grid network and controlling GHG emissions. As with any new regulatory instrument, design and implementation face significant challenges. For the REC trading market, the main obstacle comes from the loose targets that will make it difficult to create sufficient demand for RECs. Given that the RECs are generated by integrating electricity from renewable sources into the grid network, the low demand for RECs will eventually affect the penetration of renewable energy into the grid. Basing targets on the level of consumption, instead of focusing on installed capacity, is an important part of enhancing access to the grid network by renewable energy. As a relatively new regulatory mechanism, REC trading also requires a set of rules to govern its operation and, most importantly, to ensure its credibility. Among these rules, adopting penalty provisions for noncompliance is essential to encourage provinces to absorb more renewable energy rather than depending on the coal-fired generation within their geographic boundary.

For the ETS, the slow progress of the national ETS legislation has caused profound impacts on the development of China's national ETS. The ETS legislation is fundamental to all aspects of ETS operation, and it explains why the national ETS has stalled. The announcement that the national ETS was established in December 2017 was a political statement more than a real commencement of the programme. On the one hand, for industry, the major challenge is the uncertainty surrounding the national cap and allocation method, because the detailed rules are not yet available – a central question that underlies their investment strategies towards low-carbon development. On the other hand, the lack of national

87 Notice of NDRC and NEA on Establishing and Improving the Safeguard Mechanism for Renewable Energy Consumption (n 4).

88 Hove and Wetzel (n 78).

ETS legislation undermines the prospects of China's ETS pilots. In addition, the boundaries of authority and responsibility for system design and market supervision must be defined as a priority. There are important coordination opportunities and challenges that have arisen as a result of these institutional changes, which reshuffle the government agencies and which affect the progress of China's national ETS. Because more institutions are operating in the ETS space as China's ETS grows, more coordination is needed to ensure a robust legal framework governing the operation of China's ETS.

6 Regulation of sources and sinks in China's domestic offset scheme

Hao Zhang

Introduction

This chapter evaluates the institutional and regulatory framework governing the domestic offset scheme in China. It examines the substantive and procedural requirements being established to qualify the China Certified Emission Reductions (CCERs) for acceptance into China's domestic cap-and-trade schemes. The creation of China's domestic offset scheme follows on from the governance framework that the Chinese authorities established for participation in the Kyoto Protocol's Clean Development Mechanism (CDM).[1] The focus of this chapter is the regulatory landscape underpinning China's domestic offset scheme and the challenges faced by regulators to ensure its success. China's domestic offset scheme has become increasingly important due to the decline in the international trade of the CDM's Certified Emission Reductions (CERs) and the fact that China is working towards establishing one of the biggest emission-trading systems in the world.[2] China's domestic offset scheme could help underpin its national emission-trading scheme (ETS) and eventually generate emission reductions in sectors that are outside of the scope of the ETS.

China's domestic offset scheme is of particular importance in driving investment in the sectors that are vulnerable to climate change, such as agriculture and forestry.[3] China's experience may also set an example for developing countries that are striving to reduce their greenhouse gas (GHG) emissions, through incentivizing carbon-offset projects.[4] The chapter examines this domestic development

1 Alex Y. Lo and Ren Cong, 'After CDM: Domestic Carbon Offsetting in China' (2017) 141 *Journal of Cleaner Production* 1391; Hao Zhang and Christopher Arup, 'Beyond the CDM: Regulating China's Domestic Offset Scheme' (2015) 45(1) *Environmental Law Reporter* 10049.

2 Christopher Arup and Hao Zhang, 'Lessons from Regulating Carbon Offset Markets' (2015) 4(1) *Transnational Environmental Law* 69; Da Zhang et al., 'Emissions Trading in China: Progress and Prospects' (2014) 75 *Energy Policy* 9.

3 For discussion of the roles of offsets more generally, see Heather C. Lovell, 'Governing the Carbon Offset Market' (2010) 1 *Wiley Interdisciplinary Reviews: Climate Change* 353; and Ross Garnaut, *The Garnaut Climate Change Review: Final Report*, 321 (Cambridge University Press, 2008).

4 Zhongxiang Zhang, 'Carbon Emissions Trading in China: The Evolution from Pilots to a Nationwide Scheme' (2015)15 *Climate Policy* 104.

in China, with the understanding that regulation should be a learning process.[5] Such a process allows public regulators to apply lessons from a range of sources and experiences, including their engagement with international law, their knowledge of schemes in other countries, and their own governance practices.[6] In principle, this process enables regulators to assimilate expertise from various areas and enlist various stakeholders in building the domestic scheme.[7]

China's environmental governance has long been subject to a command-and-control regime, but it also needs to learn as it experiments with market mechanisms for environmental regulation.[8] It is not relieved of the commercial, environmental, and political challenges that have to be met in developing such mechanisms.[9] Currently, the Chinese domestic offset regulation is very much a work in progress.[10] Success in China in moving from participating in the CDM to a national offset scheme is critical to the development of climate law and policy in the country. An examination of the regulatory and institutional arrangements underpinning China's offset scheme suggests that there is potential for the domestic offset sector to contribute to emission reductions. However, China will have to navigate several potential pitfalls, including ensuring environmental integrity and avoiding double counting, to ensure success.

This chapter begins with China's participation in the CDM by focusing on the institutional and regulatory frameworks established in China to generate CERs. It then turns to China's domestic offset scheme. As detailed below, China has built its domestic offset scheme on the lessons learned from the CDM. This study therefore has benefited from the scholarly work undertaken on China's CDM-related regulatory and institutional arrangements.[11] The offset regulation in China is similar to the offset regulation in non-Chinese ETSs.[12] State regulation constitutes the foundation of these environmental markets.[13] The regulation structures the

5 Sanja Bogojević, *Emissions Trading Schemes: Markets, States and Law* (Hart Publishing, 2013), 174; Markus Lederer, 'Market Making via Regulation: The Role of the State in Carbon Markets' (2012) 6 *Regulation & Governance* 526; Michel Callon, 'Civilizing Markets: Carbon Trading between in vitro and in vivo Experiments' (2009) 34 *Accounting, Organizations and Society* 542.

6 Arup and Zhang (n 2).

7 Ibid.

8 Zhang (n 4); Zhang et al. (n 2).

9 Arup and Zhang (n 2).

10 Ibid.

11 Xiaoyi Jiang, *Legal Issues for Implementing the Clean Development Mechanism in China* (Springer, 2013). There is also an extensive body of literature that examines the drawbacks of offset schemes generally and that criticizes the governance framework of the CDM. See, for example, Steffen Böhm and Siddhartha Dabhi (eds), *Upsetting the Offset: The Political Economy of Carbon Markets* (MayFly Books, 2009); Tamra Gilbertson and Oscar Reyes, 'Carbon Trading: How It Works and Why It Fails' (Dag Hammarskjöld Foundation, 2009) <www.daghammarskjold.se/wp-content/uploads/2014/08/cc7_web_low.pdf> accessed 24 September 2019, chapter 4; Jim DiPeso, 'Carbon Offsets: Is the Environment Getting What You Pay For?' (2007) 17 *Environmental Quality Management* 89.

12 Arup and Zhang (n 2).

13 Lederer (n 5).

demand for offsets by specifying the share that offsets can contribute to compliance, the sources or locations from which they may be drawn, and the sectors or activities on which they may be based.[14] In China, regulations at various levels, in particular at the National Development and Reform Commission (NDRC) level and at the ETS pilot programme level, set standards for the supply and demand of offsets.

Yet the specific regulatory arrangements relating to the linkage between an offset scheme and an ETS vary considerably between jurisdictions. In China, there is an oversupply of CCERs and concerns surrounding the credibility of CCERs from a number of projects. The NDRC has suspended the approval of new projects since March 2017 in order to make necessary amendments to existing rules.[15] This raises concerns about the future of the offset market in China, which affects confidence in CCERs.[16] China's pilot ETS programmes create the major demand for CCERs, while the supply of offsets has been centralized with the NDRC.[17] The NDRC, having served as the Designated National Authority for the CDM in China, has remained influential in China's climate policy.[18]

The chapter is structured as follows. The next part looks at China's regulatory and institutional arrangements related to participation in the CDM. By examining the available studies, reports, and official documents, this part identifies where China has been successful in participating in the CDM as a supplier of CERs and also in engaging with the European Union's EU ETS to sell CERs. It reveals the capacity building, including knowledge transfer, that Chinese stakeholders have enjoyed through such processes. The subsequent part reviews China's domestic framework for regulating CCERs and provides insights into some of the regulatory challenges that the central government will need to address, particularly in the context of building the national ETS and facilitating its linkages with the domestic offset scheme in China.

Establishing China's institutional and regulatory framework governing the CERs under the CDM

As one of the Kyoto Protocol's flexibility mechanisms, the CDM's aim is to incentivize investment in emission-reduction projects in non-Annex I countries.[19]

14 Arup and Zhang (n 2).

15 Announcement of the National Development and Reform Commission, NDRC Order No. (2017)2, 14 March 2017.

16 Understanding the Initiation of the National ETS by Looking at the Suspension of the CCERs (从暂缓CCER看全国碳市场启动) (China Economic Herald, managed by the NDRC, 7 April 2017) <www.ceh.com.cn/ztbd/jnjpzk/1029753.shtml> accessed 24 September 2019.

17 Interim Measures for the Administration of Voluntary Greenhouse Gas Emission Reduction Transactions (温室气体自愿减排交易管理暂行办法), NDRC Order No. (2012)2862, 9 October 2012.

18 Measures for the Operation and Management of the Clean Development Mechanism Projects (清洁发展机制项目ff行管理办法), NDRC, Ministry of Science and Technology, Ministry of Foreign Affairs and Ministry of Finance Order No. (2011)11, 3 August 2011.

19 Kyoto Protocol to the United Nations Framework Convention on Climate Change, UN Doc FCCC/CP/1997/7/Add 1 (10 December 1997) (entered into force 16 February 2005), art. 12.

It has provided a supplementary approach for Annex I countries to meet their emission-reduction targets.[20] To date, China has hosted a substantial number of CDM projects: nearly 50% of registered projects have been located in China, and they represent around 59% of CERs issued.[21]

Under Kyoto, governments of Annex I countries may purchase CERs to meet their commitments.[22] Moreover, under the EU ETS, liable entities may purchase and surrender CERs towards compliance.[23] The CDM has thus become a channel for China's engagement with Annex I countries as well as the EU ETS.

To ensure the quality of CERs, the Kyoto Protocol and related decisions of the Conference of the Parties Serving as the Meeting of the Parties to the Kyoto Protocol (CMP) established a detailed regulatory and institutional framework. It includes modalities and procedures for the approval of CDM projects[24] and authorizes certain agencies to perform the validation, verification, and certification functions for projects and for the emission reductions they claim.[25]

As part of the framework, the CDM Executive Board (CDM EB) was established to oversee the CDM's operation.[26] The main functions of the CDM EB

20 Ibid.

21 'Distribution of Registered Projects by Host Party' (UNFCCC, 24 September 2019) <https://cdm.unfccc.int/Statistics/Public/files/201908/proj_reg_byHost.pdf> accessed 24 September 2019; 'Distribution of Expected CERs from Registered Projects by Host Party' (UNFCCC, 24 September 2019) <https://cdm.unfccc.int/Statistics/Public/files/201908/ExpRed_reg_byHost.pdf> accessed 24 September 2019.

22 Kyoto Protocol to the UNFCCC (n 19), art. 12.

23 Directive 2004/101/EC of the European Parliament and of the Council of 27 October 2004 amending Directive 2003/87/EC Establishing a Scheme for Greenhouse Gas Emission Allowance Trading within the Community, in respect of the Kyoto Protocol's Project Mechanisms [2004] OJ L338/18.

24 Decision 3/CMP.1, Modalities and Procedures for a Clean Development Mechanism as defined in Article 12 of the Kyoto Protocol, Conference of the Parties Serving as the Meeting of the Parties to the Kyoto Protocol, Report of the Conference of the Parties serving as the meeting of the Parties to the Kyoto Protocol on its First Session, held at Montreal from 28 November to 10 December 2005 – Addendum–Part 2: Action Taken by the Conference of the Parties Serving as the Meeting of the Parties to the Kyoto Protocol, UN Doc FCCC/KP/CMP/2005/5/Add.1 (30 March 2006).

25 Validation requires the DOE to conduct a thorough and independent assessment of the proposed Project Design Document against the applicable CDM requirements. This process happens after the DNA approves the project application. After validation, the proposed project can apply to register with the CDM EB. Verification is the independent review and ex-post determination by the DOE of the monitored reductions in anthropogenic emissions by sources of GHGs that have occurred as a result of a registered CDM project activity during the verification period. The DOE is to ensure that only verification activities undertaken after the publication of the monitoring report on the UNFCCC's CDM website are used as a basis for the DOE to conclude its verification. Certification is the written assurance by the DOE that during the specified period, the project activity achieved the emission reductions as verified; the DOE is then to submit a request for issuance of CERs to the CDM EB. See CDM Validation and Verification Standard (CDM, CDM-EB65-A04-STAN, version 7.0), sec. 7 and sec. 9.

26 Decision 17/CP.7, Report of the Conference of the Parties on Its Seventh Session, Held at Marrakesh from 29 October to 10 November 2001 – Addendum–Part 2: Action Taken by the Conference of the Parties, UN Doc FCCC/CP/2001/13/Add.1/Corr.1 (15 May 2002), para. 4.

include approving new project methodologies; overseeing the accreditation of Designated Operational Entities (DOEs); establishing and maintaining a database of approved rules, procedures, and methods; and developing and maintaining the CDM Registry.[27] The CDM framework also seeks to enhance quality control at the national level. Each country participating in the CDM must establish a Designated National Authority (DNA).[28] The DNA is responsible for the approval of entities participating in CDM project activities and for confirmation that CDM project activities assist the country in achieving sustainable development.[29] DOEs are another important component of CDM governance. They are independent auditors that assess whether a proposed project meets all of the eligibility requirements of the CDM (validation) and has achieved GHG-emission reductions (verification and certification).[30] They are accredited by the CDM EB to perform these functions for different sectors, depending on their expertise.[31] A DOE may be either a domestic entity or an international organization.

As a non-Annex I party to the Kyoto Protocol, China established a domestic regulatory and institutional framework in accordance with the aforementioned requirements. The main regulation governing CDM project activities is the Measures for the Operation and Management of the Clean Development Mechanism Projects, promulgated by the NDRC in 2011.[32] The Measures define the rights and obligations of key entities that are involved in carrying out CDM project activities in China. These entities consist of the CDM management institutions, eligible project participants, and the DOEs.[33] The CDM management institutions are China's CDM Review Board and the NDRC.[34] Eligible project participants in China include Chinese-funded or Chinese-controlled enterprises located inside China.[35] These enterprises are the project owners; they may cooperate with foreign parties on CDM projects.[36] The DOEs are the verification and certification institutions that are accredited by the CDM EB to operate in China.[37]

27 Decision 3/CMP.1 (n 24).
28 Ibid.
29 Ibid.
30 For a full list of DOE functions, see Decision 3/CMP.1, and annex to the Decision 3/CMP.1 (n 24), para. 27.
31 Decision 3/CMP.1 (n 24).
32 Measures for the Operation and Management of the Clean Development Mechanism Projects (n 18).
33 Ibid.
34 Ibid., arts. 11 and 12.
35 Ibid., art. 10.
36 For CDM projects, the restriction on foreign investment is strict that the project developer or the project owner needs to have majority control of the enterprise. As discussed below, for China's domestic offset scheme there is no such restriction on foreign investment as long as the project participant is an entity registered in China.
37 Measures for the Operation and Management of the Clean Development Mechanism Projects (n 18), art. 35.

A significant feature of the management institutions is the establishment of China's CDM Review Board. The Board is a leading group, chaired by the NDRC and the Ministry of Science and Technology, vice chaired by the Ministry of Foreign Affairs, and comprised of several other ministries and bureaus as its members.[38] The Board examines important issues affecting the project application process, such as the methods to be applied, the price of CERs transacted, whether funding for emission reductions is additional to business as usual, the level of technology transfer, and the projects' anticipated effects in enhancing sustainable development.[39]

The NDRC, serving as China's DNA under the CDM, is principally responsible for managing the project application process and approving CDM projects in China. The NDRC also arranges for the expert review of applications for participation in CDM project development.[40] The NDRC is centre stage in CDM management in China.[41] Its regulatory position has enabled it to play a key role in influencing investment in project development.[42] The regulations and rules adopted by the NDRC not only stipulate the substantive and procedural requirements for developing CDM projects in China; they also convey crucial policy settings that shape the various investors' approaches towards the CDM, in particular their choice of certain types of project activities.[43]

China's experience with the CDM in the domestic regulatory environment demonstrates how it has benefited certain sectors.[44] The industrial gas, hydroelectricity, and wind-power sectors became the dominant CDM project types in China.[45]

China's domestic framework governing CCERs

Substantive requirements

The national regulations that govern the domestic offset scheme in China consist of two essential documents. The first is the Interim Measures for the

38 Ibid., art. 8.

39 Ibid., art. 11.

40 Ibid., art. 12.

41 Sebastian Thomas, Paul Dargusch and Andrew Griffiths, 'The Drivers and Outcomes of the Clean Development Mechanism in China' (2011) 21 *Environmental Policy and Governance* 223.

42 Ibid.

43 Measures for the Operation and Management of the Clean Development Mechanism Projects (n 18), art. 36. Without ruling out any sector from the eligible CDM projects, the NDRC has prioritized certain sectors through determining their transaction costs by using taxation policy. In general, the Chinese government agency taxes priority project at 2% of the value of CERs and much higher for "low-hanging-fruit" projects (65% for HFC projects, 30% for N_2O projects that produce adipic acid, 10% for N_2O projects that produce nitric acid, and 5% for PFC projects). These arrangements affect cost–benefit analyses and thus influence the decision-making process of project developers.

44 Thomas, Dargusch and Griffiths (n 41).

45 Ibid., the authors conducted a study of 579 registered CDM projects in China and found that the benefits were primarily economic rather than environmental.

Administration of Voluntary Greenhouse Gas Emission Reduction Transactions (Interim Measures),[46] which sets up a legal framework for approval and participation by projects developing CCERs. The second is the Guidelines on Validation and Verification for Voluntary Greenhouse Gas Emission Reduction Projects.[47] It specifies the accreditation criteria and procedures for independent third-party verifiers. It also clarifies the requirements for validation and verification services that the third-party verifiers are to provide.

The Interim Measures stipulate the substantive and procedural requirements for the production of CCERs. The substantive requirements contain several important components that are essential to guaranteeing that CCERs represent genuine and environmentally sustainable reductions. They include the types of GHGs covered by CCERs; the eligibility of project participants; methods for project development; and validation, verification, and certification arrangements.

The Interim Measures specify the six types of GHG[48] from which CCERs may be generated. Compared to the CDM's requirements, a wider range of entities are eligible to participate in CCER development. As noted earlier, legal entities registered in China are eligible to participate in the domestic offset scheme.[49] Compared to the conditions for CDM projects, which require project participants to be Chinese-funded or Chinese-controlled enterprises (excluding foreign enterprises), there is no restriction on foreign investment in China's domestic offset scheme, meaning that the potential number of participants will be larger than the CDM.[50] The project-eligibility requirements contain both the time and the activity criteria, namely that implementation started after 16 February 2005 and the new project is developed on the basis of an approved method.[51] Certain China-based CDM projects are eligible to be converted into domestic-scheme offset projects, although there are restrictions, principally to avoid double counting reductions (see discussion below). The criteria for qualifying these projects are that a project has been approved by the NDRC as a CDM project but has not yet been registered under the CDM or any other emission-reduction scheme; that the project is a registered CDM project with emission reductions

46 Interim Measures for the Administration of Voluntary Greenhouse Gas Emission Reduction Transactions (n 17).

47 Guidelines on Validation and Verification for Voluntary Greenhouse Gas Emission Reduction Projects (温室气体自愿减排项目审定与核证指南), NDRC Order No. (2012)2862, 9 October 2012.

48 Interim Measures for the Administration of Voluntary Greenhouse Gas Emission Reduction Transactions (n 17), art. 2.

49 Ibid., art. 5. The domestic and foreign institutions, enterprises, organizations, and individuals are entitled to participate in voluntary GHG emission reduction transactions.

50 For the eligibility of CDM participants, see Measures for the Operation and Management of the Clean Development Mechanism Projects (n 18), art. 10. This provision has a particular restriction on foreign investments: only Chinese-funded or Chinese-controlled enterprises are eligible for the CDM project activities.

51 Interim Measures for the Administration of Voluntary Greenhouse Gas Emission Reduction Transactions (n 17), art. 13.

occurring before the registration date; or that the project is a registered CDM project with no CERs issued.[52]

Similar to the CDM, the Interim Measures require emission reductions generated by a project to be calculated on the basis of an approved method.[53] Newly developed methods, on the basis of experience with the CDM, may be filed with the NDRC, along with the design documents for such methods and the relevant projects.[54] The Interim Measures allow appropriate methods approved by the CDM EB to be applied in the domestic offset projects to serve the purpose of determining the project baseline, authenticating the additionality of a project activity, calculating the emission reductions, and preparing monitoring plans to oversee project implementation.[55]

The Interim Measures place no categorical restriction on the type of activity or sector from which CCERs may be sourced. The NDRC has published 200 approved methods online.[56] Of these, 173 have been drawn from the CDM methods. This enables the transfer of learning about standards and guidelines that have been tested in practice in China. So far, there are five non-CDM methods on land use, land-use change, and forestry (LULUCF) activities.[57] These methods were developed to incentivize carbon sequestration in the land and forestry sector. The forestry sector has been given particular attention in all national policies and actions addressing climate change due to its potential as a carbon sink and the central government's commitment to increase forest coverage.[58] In addition to the LULUCF initiative, the list of methods reflects China's traditional focus on renewable energy, energy efficiency, fuel switch, and methane-gas reduction,[59] although it also includes methods for hydrofluorocarbon (HFC) and N_2O industrial-gas offsets, which the EU ETS banned after 2012 in response to criticism of their environmental integrity.[60]

Third-party agencies accredited by the NDRC validate project components and confirm the emission reductions generated.[61] To safeguard the credibility of offset

52 Ibid., art. 13.
53 Ibid., art. 9.
54 Ibid., art. 11.
55 Ibid., art. 10.
56 The list of methodologies is available at <http://cdm.ccchina.org.cn/zylist.aspx?clmId=162> accessed 24 September 2019.
57 Ibid.
58 'China's Policies and Actions for Addressing Climate Change' (Ministry of Ecology and Environment, November 2018) 15–16 <http://english.mee.gov.cn/News_service/news_release/201812/P020181203536441502157.pdf> accessed 29 October 2019.
59 Jiang (n 11); Thomas, Dargusch, and Griffiths (n 41).
60 These industrial gas offsets were excluded from 1 May 2013. See Commission Regulation (EU) No 550/2011 of 7 June 2011 on Determining, Pursuant to Directive 2003/87/EC of the European Parliament and of the Council, Certain Restrictions Applicable to the Use of International Credits from Projects involving Industrial Gases [2011] OJ L149/1.
61 Interim Measures for the Administration of Voluntary Greenhouse Gas Emission Reduction Transactions (n 17), arts. 12 and 18.

credits, the Interim Measures require that an agency that has validated a project with an annual emission reduction of 60,000 tonnes or more not be permitted to verify emission reductions for the same project.[62]

Institutional and procedural requirements

As with the substantive requirements, the NDRC has modelled the institutional and procedural framework underpinning China's domestic offset scheme on the CDM, with some adjustments made to streamline the process. Working principally on the model of the procedures for CDM project development, the NDRC's responsibility is to ensure quality control at different stages of the project-development cycle – as shown in Table 6.1. The quality control pertains to the procedures on reviewing project applications, checking project eligibility for participation, and reviewing the verification and certification of a project before CCERs are issued.[63]

Due to the large number of applications foreseen, the application procedures for some project developers are devolved to local levels.[64] This helps to relieve the NDRC's workload, but it may also raise an issue of coordination between the NDRC and its provincial counterparts in encouraging investment in offset sectors. Due to the different policy agendas and other interests of provincial governments, the number of offset projects can be expected to vary significantly among them, also taking into account the divergent requirements for accepting CCERs in China's ETS pilot programmes.[65]

Some state-owned enterprises (SOEs) supervised by the State-Owned Assets Supervision and Administration Commission have been granted permission to apply directly to the NDRC for approval of offset projects. A number of SOEs are encouraged to participate in CCER projects, in light of their need to increase energy efficiency and reduce emissions.[66] So far, there are 43 SOEs that are qualified to initiate direct applications.[67] Project developers, other than the 43 SOEs, are required to submit an application to the Development and Reform Commission of the province, autonomous region, or municipality where the proposed project is located.[68]

62 Ibid., art. 18.

63 Interim Measures for the Administration of Voluntary Greenhouse Gas Emission Reduction Transactions (n 17).

64 Ibid., art. 14, para. 2.

65 See below part 4.2 for further discussion.

66 Benoit Mayer, Mikko Rajavuori, and Fang Meng, 'The Contribution of State-Owned Enterprises to Climate Change Mitigation in China' (2017) 7(2–3) *Climate Law* 97.

67 Interim Measures for the Administration of Voluntary Greenhouse Gas Emission Reduction Transactions (n 17), Annex I. Annex I contains a list of enterprises, supervised under the State-Owned Assets Supervision and Administration Commission, that are able to lodge a direct application to the NDRC.

68 Ibid., art. 14, para. 2.

Table 6.1 Procedural requirements for generating CERs in China and streamlined process for developing CCERs

Procedural Requirements	*Application*	*Approval*	*Registration*	*Issuance*
China CERs* (**Step**, Actor, *Period*)	**Submission**, Project Owner; **Expert Review**, NDRC, *Within 30 days*	**Approval**, NDRC, *Within 20 days*; **Validation**, DOE, *Within 30 days*	**Registration**, CDM EB, *Within 8 weeks*; **Report**, Project Owner, *Within 10 days*; **Implementation**, Project Developer	**Verification and Certification**, DOE; **CER Issuance**, CDM EB, *Within 15 days*; **CER Record**, NDRC
China CCERs** (**Step**, Actor, *Period*)	**Submission**, Project Owner; **Expert Review**, NDRC, *Within 30 days*	**Approval**, NDRC, *Within 30 days*; **Validation**, Accredited Verifier, *Within 30 days*; **Implementation**, Project Developer		**Verification and Certification**, Accredited Verifier; **Expert Review**, NDRC, *Within 30 days*; **CCER Issuance**, NDRC, *Within 30 days*; **CCER Record**, NDRC

Source: *Jiang (n 11), 97. **Compiled on the basis of the Interim Measures for the Administration of Voluntary Greenhouse Gas Emission Reduction Transactions (n 17).

A comparison of the procedures that underpin CERs with those for CCERs reveals that the key difference between them is the lack of a registration requirement for CCER developers (Table 6.1). When designing the domestic offset scheme, the NDRC took account of the CDM's lengthy and costly registration process.[69] It was a key factor preventing some sectors in China from participating in the CDM,[70] especially for projects in the LULUCF category.[71] The production of CCERs will thus be faster and less expensive.[72] The potential risk, as discussed

69 Shu Wang and Maosheng Duan, 'China's Domestic Offset Scheme' (Partnership for Market Readiness, World Bank, 2 March 2014) <www.thepmr.org/system/files/documents/China%27s%20domestic%20offset%20scheme%20-20140226_0.pdf> accessed 24 September 2019.

70 Ibid.

71 Zhijun Zhang et al., 'Costs of Carbon Sequestration for Afforestation/Reforestation Projects Under Clean Development Mechanism: A Case Study of Pearl River Basin in Guangxi, China (清洁发展机制（CDM）造林再造林项目碳汇成本研究--以CDM广西珠江流域治理再造林项目为例)' (2009) 5 *Climate Change Research* (气候变化研究进展) 348.

72 As observed by Hoogzaad and Streck, "registration of a project takes on average more than 400 days and may even take as long as 700 days in the case of a CDM project that requires approval of

below, is that the NDRC will face a difficult task in keeping track of the supply and demand of CCERs, particularly when the supply of CCERs far exceeds demand. The NDRC will also face a challenge from the industrial-gas projects that it has allowed to be part of the domestic offset scheme. If the large quantity of credits generated from CDM industrial-gas projects is any guide, they will profoundly impact supply in the domestic offset market.

Linkage between CCERs and China's ETS programmes

The success of CCERs in directing investment into the offset sectors in China is ultimately driven by the price of CCERs, compared with the price of allowances in the ETS pilot programmes. The question is whether the current decision-making framework has the capacity to generate supply and demand for CCERs at a predictable and substantial price. By July 2019, the total number of CCERs purchased by liable entities under the ETS pilots reached 194 million tons of CO_2, and the CCER price varied among ETS pilots, with around RMB20 per ton in Shanghai and RMB4–RMB20 per ton in Beijing.[73] Liable entities in Shanghai purchased CCERs totaling 84.91 million tons, followed by Guangdong, Beijing, and Shenzhen (10–40 million tons).[74] Hubei and Tianjin recorded lesser trading volumes (<10 million tons).[75] Given the slow progress of the national ETS (see Chapter 5), stakeholders eagerly await the release of detailed arrangements on the linkage between the national ETS and the domestic offset scheme.[76] For the time being, the demand for CCERs is almost entirely driven by the ETS pilot programmes. For the ETS pilot programmes, both offset producers and liable entities need to plan ahead if they are to trade CCERs for compliance purposes.[77]

Under the current framework, the NDRC plays a crucial role in controlling both the quantity and quality of CCERs. Yet at the same time, the local authorities at the pilot programmes have the responsibility to decide the volume, sectors, and

a new methodology", and project developers face uncertainties "on a future revenue stream that is crucial to the feasibility of the project". Jelmer Hoogzaad and Charlotte Streck, 'A Mechanism with a Bright Future: Joint Implementation', in David Freestone and Charlotte Streck (eds), *Legal Aspects of Carbon Trading: Kyoto, Copenhagen, and Beyond*, 176–194 (Oxford University Press, 2009).

73 Other ETS pilots rarely publish the price of CCERs. 'Weekly Report of Carbon Market in China' (8–12 July 2019) (Tanjiaoyi website, 2019) <www.tanjiaoyi.com/article-27927-1.html> accessed 29 October 2019.

74 Ibid.

75 Ibid.

76 Huw Slater et al., '2018 China Carbon Pricing Survey' (China Carbon Forum, 2018) <www.eu-chinaets.org/upload/file/20180906/1536164718942317.pdf> accessed 29 October 2019.

77 At least for now, the trading platforms, such as the Beijing Environmental Exchange, are acting as intermediaries that strive to connect the suppliers of CCERs with the buyers under each pilot programme. See, for example, Notice on Establishing Contacts between the Sellers and Buyers with the Intention to Transact CCERs (核证自愿减排量意向交易方征集活动的通知) (Beijing Environmental Exchange, 2014) <www.bjets.com.cn/article/xxzx/jysgg/201403/20140300000159.shtml> accessed 24 September 2019.

sources of the credits that are allowed to enter their schemes. The domestic offset scheme therefore needs to harmonize the efforts of the multiple stakeholders. The localized decision-making process, on the demand side of CCERs, has raised concerns about the likely environmental integrity of the offsets.[78] This is due to the variety of offset regulations in the pilot programmes with different preferences for certain types of CCERs, such as those produced locally by industrial-gas projects.[79] This tendency towards local protectionism is a recurring theme in China, arising from its governance structure and decentralized framework in implementing environmental law.[80]

The NDRC is acutely aware of the importance of environmental integrity and the danger that, if it opens up the market too quickly, it could have a situation of oversupply in which the value of offset credits drops precipitously.[81] However, project developers in China are aware of the many eligible projects that could be converted from the CDM to the domestic offset market; by end of April 2018, there were 2,871 projects put forth for review, and 1,047 of them had been registered.[82] As to the issuance of credits, the CCERs from 254 of the registered projects were publicly available, and more CCERs are in the pipeline from the rest of the registered projects. Among the 254 projects with certification reports, which represent 52.94 million tons of CCERs, 64.2% of the CCERs are conversion projects from the China-based CDM. Practically, the conversion will enable a large number of CCERs to flood the market and thus drive the price of CCERs to extremely low levels. Regarding the volume of offsets, the regulations underpinning the pilot programmes allow the liable entities to discharge around 5%–10% of their obligations by surrendering CCERs. For instance, the acceptance percentage in Shanghai and Guangdong is 5% and 10% respectively.[83]

Moreover, the local authorities governing the ETS pilot programmes also limit the sectors and sources from which the CCERs are to be generated.[84] For example,

78 Xiaojian Zhong et al., 'The Features, Motivations and Tread of China's Carbon Trading Program: A Comparative Study based on International Experiences' (中国碳交易市场的特征、动力机制与趋势 – 基于国际经验比较) (2017) 37 *Acta Ecologica Sinica* (生态学报) 331.

79 Arup and Zhang (n 2); Zhang et al. (n 2).

80 Michael Gillenwater and Stephen Seres, 'The CDM: A Review of the First International Offset Program' (2011) 3–4 *Greenhouse Gas Measurement & Management* 179.

81 Daphne Yin, 'China's Carbon Emissions Traders Await Offset Demand' (Ecosystem Marketplace, 9 July 2013) <www.ecosystemmarketplace.com/pages/dynamic/article.page.php?page_id=9820> accessed 24 September 2019.

82 Slater et al. (n 76).

83 Implementation Measures for Allocating and Managing Emission Allowances in Shanghai (上海市2013–2015碳排放配额分配和管理方案), Shanghai Development and Reform Commission Order No. (2013)168, 22 November 2013, art. 4; Interim Measures for Managing and Trading Emissions Allowances in Guangdong Province (广东省碳排放管理试行办法), Guangdong People's Government Order No. (2014)197, 15 January 2014, art. 19.

84 Zhiwen Gao, 'Analysis of CCER Restrictive Policies in China's Carbon Emission Trading Pilots' (中国各碳排放权交易试点CCER限制性政策分析) (Climate Bridge, 2014) <www.climatebridge.com/news/DownLoad.aspx?ID=/uploads/20150918/20150918160948367.pdf> accessed 24 September 2019.

the Hubei pilot programme allows only CCERs from Hubei province to be used for compliance by liable entities, with no further limitation on sectors and project types.[85] The Shenzhen pilot programme excludes local CCERs that come from the sectors that are covered by the caps, to avoid the issue of double counting.[86]

The biggest demand for CCERs comes from the two provinces that favour locally generated offsets. Depending on the size of the cap, which may change slightly across different compliance periods, Hubei's liable entities may need up to around 23 Mt CO_{2}-eq of CCERs in 2018 from the province for compliance purposes.[87] The Guangdong regulation requires that 70% of offset be sourced locally, meaning that as much as 28 Mt CO_{2}-eq of local CCERs was in demand in 2018.[88] All seven pilot programmes accept CCERs from the LULUCF category for compliance. The discrimination in some programmes against offsets that are not local has caused concerns that supply will exceed demand in other pilot programmes that are more open to CCERs from outside.[89] The demand from other pilot programmes is much smaller for other reasons too, such as a smaller cap and a more restricted volume of offsets for compliance.

The positive development so far is that a key factor in selecting the offsets to be linked to the local ETS is the environmental quality of the emission reductions. For the pilot programmes, the CCERs retain their attraction as long as the selection and standard-setting ensure environmental sustainability, although setting such standards places great demands on the NDRC's regulatory capacity. Clear direction from the central level is a means to consolidating CDM expertise in verification and then introducing this expertise into standard-setting bodies that operate at the local level.[90]

Conclusion

Overall, the experiences from the ETS pilot programmes have provided some important lessons for the future of the domestic offset market in China, particularly

85 Interim Measures for Managing and Trading Emissions Allowances in Hubei Province (湖北碳排放权管理和交易暂行办法), Hubei People's Government Order No. (2014)371, 23 April 2014, art. 18.

86 Interim Measures for Managing and Trading Emissions Allowances in the City of Shenzhen (深圳市碳排放权交易管理暂行办法), Shenzhen People's Government Order No. (2014)262, 19 March 2014, art. 37.

87 Allocation Plan for Allowances under Hubei's Emission Trading Pilot Program in 2018 (湖北省2018年碳排放权配额分配方案), Department of Ecology and Environment of Hubei Provincial Government, 10 July 2019. This is a rough calculation that is based on the total cap, given that the total number of allowances allocated to the liable entities is unavailable in the allocation plan.

88 Similarly, the calculation is conducted on the basis of information available in the allocation plan in Guangdong. Implementation Plan for Allowance Allocation in Guangdong in 2018 (广东省2018年度碳排放配额分配实施方案), Guangdong Development and Reform Commission, 26 July 2018.

89 Yin (n 81).

90 The development of the hybrid public-private standard-setting bodies originated from the infrastructure and capacity building that has been achieved through participation in the CDM. The private third-party verifiers have been playing an important role in developing methods, and the public body has been active in administering project development. See Jiang (n 11), chapter 3.

in the context of developing the national ETS. First, the streamlined procedure to generate CCERs attracts potential developers, but a speedy process of project development also requires stronger oversight of the environmental sustainability of the CCERs. In this regard, the NDRC has a stronger role to play in managing both the procedural and substantive aspects governing the offset project activities in China.

Furthermore, the CCERs are subject to the overall performance of the ETS itself. For instance, a loose cap leading to the over-allocation of allowances under the ETS will result in a chain reaction, making CCERs much less attractive to liable entities. The reduction of market viability for CCERs will eventually affect investments in carbon-offset and sequestration projects, which are integral parts of the low-carbon investment portfolio in China.

Given China's importance in the global efforts to mitigate climate change, its domestic actions towards emission reductions will have both domestic and international implications. Because China is working progressively to construct a national ETS through local pilot programmes, a functioning domestic offset programme is an essential mechanism to allow liable entities to explore cost-effective measures to meet compliance targets. The importance of linking the offset mechanism with the ETS in China is to provide financial incentives to sectors (such as forestry and land) that are traditionally unable to attract low-carbon investment.

However, the challenges in relation to ensuring environmental integrity and maintaining a reasonable level of supply of offset credits are ever-present for the Chinese authorities. At the international level, China's participation in the CDM and its engagement with the EU ETS via the CDM have decreased substantially. On the domestic side, the experiences so far from China's ETS pilots reveal that there are regulatory tensions between the responsible government agencies at both central and local levels as well as between local regulators who can decide the level of acceptance for CCERs.

The first tension exists in the dynamics of coordination between the NDRC and the local authorities that operate the ETS pilot programmes. The current decision-making framework has difficulties securing the environmental integrity of the scheme and driving investments into the prioritized areas, because the decision-making authority regarding supply and demand of CCERs is split between central and local authorities. The second tension is more from a compliance point of view under the ETS pilot programmes in China. The issue of exclusions and inclusions between the pilot programmes indicates that there is insufficient coordination between local authorities that regulate the shares, sources, sectors, and standards of offset credits that are eligible for compliance purposes under different pilot programmes.

A solution to resolve the tensions is that the higher-level administration (the NDRC) needs to be more proactive in guiding local authorities in their use of offset credits. This is of particular relevance under the national ETS, where the NDRC is to set ground rules about the structure of demand, such as the maximum shares of offset credits that can be used and the sectors that are prioritized to meet compliance purposes. Most importantly, the NDRC needs to review the existing regulatory framework and make changes to the permitted methodologies in order to ensure that the offset scheme is environmental sustainable. The issue of trading CCERs is also highly relevant because it relates to the commercial viability of the offset market overall. To gauge the supply and demand of CCERs, the NDRC needs to have better oversight of the market and its performance. Essentially, these ground rules will need to be taken into consideration when designing the national ETS.

The success of China in moving from participation in the CDM to a national offset scheme is a critical part of the development of climate law and policy in China. An examination of the regulatory and institutional arrangements underpinning China's offset scheme suggests that there is great potential for the domestic offset sector to contribute to emission reductions. Addressing the challenges identified in this chapter, moreover, is essential to the success of China's domestic offset scheme and the national ETS. The success of a national offset scheme will enrich the policy options and governance capacity for achieving reductions in China's emissions. China's experience of moving beyond the CDM to establish the domestic offset scheme also provides some useful insights to other developing countries that face obstacles to drive investment into sectors that are vulnerable to climate change.

7 Monitoring, reporting, and verification of greenhouse gas emissions in China

A legal analysis

Alexander Zahar and Hao Zhang

Introduction

The Conference of the Parties (COP) to the United Nations Framework Convention on Climate Change (UNFCCC) held in Bali in 2007 marked an important turning point in the history of the international climate change negotiations. Developing countries, including China, agreed at that COP to consider adopting measurable, reportable and verifiable (MRV) mitigation actions under the Convention.[1] The Bali COP's headline news was this new readiness of developing countries to formally endorse mitigation actions, which until that point had been seen as a responsibility resting squarely on developed countries. The less newsworthy item, but a no less important one – and the one that provides the international legal background to this chapter – was that mitigation plans by developing countries would be paralleled by the establishment of national MRV systems in those countries in order to facilitate international reporting on the implementation of their plans.

By 2019, all developed countries (or, to be precise, all Annex I parties to the UNFCCC) possessed high-quality MRV systems for greenhouse gas emissions (and removals[2]) as a result of their differentiated and more arduous obligations under the UNFCCC. Each had a near-perfect inventory of its emissions for every year since 1990. By contrast, few developing countries had managed to construct MRV systems that even approached the Annex I level of sophistication. In fact, most developing countries did not have such a system at all, certainly not a permanent one. In this chapter, we try to understand where China's MRV system for greenhouse gases lies along this spectrum and the pace at which it is being improved.

As the biggest energy consumer and largest greenhouse gas emitter,[3] China is planning to move towards a low-carbon economy to reduce its reliance on fossil

1 COP (UNFCCC), *Decision 1/CP.13, Bali Action Plan*, FCCC/CP/2007/6/Add.1 (2007), para. 1.

2 For brevity, in this chapter, *MRV of greenhouse gas emissions* includes MRV of greenhouse gas removals.

3 International Energy Agency, 'Global Energy and CO_2 Status Report 2018: The Latest Trends in Energy and Emissions in 2018' <https://webstore.iea.org/global-energy-co2-status-report-2018> accessed 3 December 2019.

fuels and to increase the share of renewable energy in its overall energy structure. Among the key regulatory and governance frameworks that China needs to achieve its mitigation goals, an MRV system for greenhouse gases is of particular importance. Accurate and complete data on greenhouse gas emissions are necessary both domestically – for China to design and gain feedback on its emission-reduction policies, for example – and internationally – to inform the international community on its actions on climate change, not only under the UNFCCC but also under the 2015 Paris Agreement. The latter also requires each country, including China, to report internationally on its progress against its nationally determined contribution (NDC).[4]

This chapter's critical account of China's MRV system for greenhouse gases is, to our knowledge, unprecedented. While an understanding of the development of China's greenhouse gas MRV system is valuable in itself, it is also relevant to understanding the challenges faced by other developing countries that are in the process of improving their performance under the UNFCCC and responding to the new imperatives of the Paris Agreement on the establishment of MRV systems.

MRV systems defined

Any action may be subject to MRV. In general, MRV systems, by collecting performance data, function to support policy development by giving policymakers feedback on progress towards their goals.[5] An MRV system for greenhouse gas emissions is a legal requirement in international law (as explained below), and it may also be required by domestic law. Examples of both will be given later in this chapter.

Concerning climate change actions by states, systems have been developed for MRV of policies and measures, as well as climate finance, technology transfer, and so on,[6] but these are not the main concern of this chapter. We will focus instead on MRV systems for greenhouse gas emissions, both at the international level and in China's domestic context specifically. This is not to say that MRV systems relating to other subjects regulated under the international climate treaties are not important. They are, in fact, highly important for full accountability regarding state action on climate change. Nevertheless, a system for the MRV of greenhouse gases is more fundamental, because the other systems draw on this foundational information, and therefore its proper functioning is a priority matter.

4 Paris Agreement, Article 4(13).

5 Angela Falconer et al., *Tracking Emissions and Mitigation Actions: Evaluation of MRV Systems in China, Germany, Italy, and the United States* (San Francisco, Climate Policy Initiative, 2012), i.

6 Esther Turnhout, Margaret M. Skutsch, and Jessica de Koning, 'Carbon Accounting', in Karin Bäckstrand and Eva Lövbrand (eds), *Research Handbook on Climate Governance*, 366–376, 369 (Edward Elgar, 2015).

Elements of the basic concept

We begin by defining the constituent elements of monitoring, reporting, and verification (MRV).[7] *Monitoring* includes the notion of measurement: "The term 'monitoring' summarizes various screening operations and actions for gathering information about whether the regulatees comply with environmental regulation, that is, whether they modify their behaviour in accordance with the legal requirements."[8]

Greenhouse gas monitoring involves estimating emissions in a standardized way by expressing the physical measurement of emissions, or estimates of emissions obtained indirectly through calculations, in terms of a common unit, normally a ton of carbon dioxide equivalent (1t CO_2 eq.).[9]

It is necessary to distinguish between monitoring at the domestic level and monitoring at the international level. At the domestic level, there are two main approaches to monitoring. One is to require different classes of greenhouse gas emitters (normally starting with electricity generators and energy-intensive enterprises above a certain size) to monitor their activities and file reports with the government on a regular basis. Such forms of self-monitoring are widely used in many countries, including China, and the government regulators substantially rely on the information received in this way. Self-monitoring can be enhanced through independent, private-sector verification of reports.

The second approach differs in its management of verification (as opposed to monitoring and reporting). Under this approach, the regulator carries out inspection visits to the facilities that are required to self-monitor. Inspection procedures normally include an evaluation by the inspectors of the data gathered by the facilities during the process of self-monitoring, as well as the taking of samples of emissions (or of emission factors and activity data in the case of inspections relating to greenhouse gas emissions) at the inspected facility.[10] China's approach to environmental regulation is, in general, relatively inspection-intensive.[11]

At the international level, monitoring greenhouse gas emissions is performed by state governments. Specialized government agencies, whether standing or ad hoc, collect and aggregate the results of domestic monitoring. They also, where necessary, fill in the gaps in the national inventory by calculating the emissions from economic sectors for which there is no emission monitoring imposed by regulation, such as in agriculture or forestry, by using statistical or satellite (or other) information and specialized formulae. The international organization receiving

7 For a general introduction, see Neelam Singh, Jared Finnegan, and Kelly Levin, *MRV 101: Understanding Measurement, Reporting, and Verification of Climate Change Mitigation*, 4–5 (World Resources Institute, 2016).

8 Jale Tosun, 'Environmental Monitoring and Enforcement in Europe: A Review of Empirical Research' (2012) 22 *Environmental Policy and Governance* 437, 438–439.

9 Turnhout, Skutsch, and de Koning (n 6), 366.

10 Tosun (n 8), 438–439.

11 Alexander Zahar, 'Implementation of the Polluter Pays Principle in China' (2018) 27(3) *Review of European, Comparative and International Environmental Law* 293.

state reports on greenhouse gas emissions – the UNFCCC Secretariat – arranges for the verification of a certain class of these reports by expert review teams (as discussed below).

MRV as foundational for policy

MRV is a necessary adjunct to mitigation policy, because in its absence, it would be difficult to know with any certainty whether an emission-reduction policy is working as intended.[12] It has been said that "carbon accounting operates as a technology of climate governance", because it enables outside scrutiny of state or industry actions, which increases accountability.[13] But in addition to accountability, the creation and communication of accurate information on greenhouse gas emissions can be a condition sine qua non for certain mitigation policies, such as emission trading.[14] The trading of emission allowances would not be possible if participants did not trust the trading mechanism's MRV system.

At the international level, MRV is needed to understand, among other things, the sources and trends in anthropogenic emissions, whether international declarations by states to reduce their emissions are being kept, and whether the burdens of the response to climate change are being shared equitably among states.

MRV as a component of "transparency" theory

MRV has been theorized as being valuable beyond its primary purpose of providing reliable information to policymakers. In the climate change context, the MRV of state greenhouse gas emissions is supported by a theory that a global, rule-based approach to MRV contributes to realizing the higher value of "transparency" – a foundation for the attainment of even higher values. According to this theory, transparency in state reporting on greenhouse gas emissions is needed to improve mutual "trust" among states. In this conception, increased trust is a precondition for raising the "ambition" of states' contributions to mitigation and to adaptation and financial and capacity-building support to other states, under the international climate change treaties.

12 Singh, Finnegan, and Levin (n 7), 1 ("Measurement is needed to identify emissions trends, determine where to focus greenhouse gas reduction efforts, track mitigation-related support, assess whether mitigation actions . . . are proving effective, evaluate the impact of support received, and monitor progress achieved in reducing emissions").

13 Turnhout, Skutsch, and de Koning (n 6), 372. See also Aarti Gupta and Michael Mason, 'Transparency', in Karin Bäckstrand and Eva Lövbrand (eds), *Research Handbook on Climate Governance*, 446–457, 448 (Cheltenham, UK: Edward Elgar, 2015).

14 Daniel H. Cole, 'Origins of Emissions Trading in Theory and Early Practice', in Stefan E. Weishaar (ed), *Research Handbook on Emissions Trading*, 9–26, 11, 13 (Edward Elgar, 2016); Renhu Tang et al., 'Key Challenges for the Establishment of the Monitoring, Reporting and Verification (MRV) System in China's National Carbon Emissions Trading Market' (2018) 18(S1) *Climate Policy* S106, S107; and High-Level Commission on Carbon Prices, *Report of the High-Level Commission on Carbon Prices*, 13 (Washington, DC, World Bank, 2017).

The official UNFCCC position, as expressed by the UNFCCC's Secretariat in the following passage, not only reflects the foregoing theory but also asserts that MRV capacity is crucial to how quickly and effectively a country can address climate change:

> Through the submission of national reports, their review and multilateral consideration processes under the existing measurement, reporting and verification [MRV] system under the Convention, the progress made by each developed country for the pre-2020 period is tracked. The current measurement, reporting and verification system under the UNFCCC provides a solid foundation for the enhanced transparency framework under [Article 13 of] the Paris Agreement. This enhanced transparency framework serves as the key enabler for raising ambition of climate action by all Parties under the Paris Agreement, together with the nationally determined contribution submissions and global stocktake. Transparency is also vital in building mutual trust and confidence among Parties and in promoting accountability by Parties.[15]

The assumptions underlying this theory find some support in the scholarly literature.[16] However, there is also scepticism about relying too heavily on "transparency" to achieve higher goals.[17]

Another issue is that, although at times MRV is presented as an uncontroversial, mundane activity that all states can agree to perform – at least on the condition that sufficient resources are made available to them – the MRV requirement, especially the elements of reporting and verification, impinge on state sovereignty to some extent and have not always been treated as an entirely neutral exercise.[18]

15 Katia Simeonova, *Letter to Expert Reviewers* (on file with the authors) (UNFCCC, 30 May 2019). See also <https://unfccc.int/process-and-meetings/transparency-and-reporting/reporting-and-review-under-the-paris-agreement/reporting-and-review-under-the-paris-agreement> ("The enhanced transparency framework represents an important component of the ambition cycle in the global climate regime established by the Paris Agreement by building trust and confidence that countries are taking action to meet their national climate targets and actions defined in their NDCs under the Paris Agreement.")

16 See, for example, Gupta and Mason (n 13), 446; Singh, Finnegan, and Levin (n 7), 1 and 4; Felicity Deane, Evan Hamman, and Yilin Pei, 'Principles of Transparency in Emissions Trading Schemes: The Chinese Experience' (2017) 6(1) *Transnational Environmental Law* 87, 90 and 106; and Taryn Fransen et al., *Enhancing NDCs: A Guide to Strengthening National Climate Plans by 2020*, 4 (Washington, DC: World Resources Institute, 2019).

17 See, for example, Amitai Etzioni, 'Is Transparency the Best Disinfectant?' (2010) 18(4) *The Journal of Political Philosophy* 389, 395–396 ("There continues to be a dearth of studies empirically testing the theoretical claims of transparency advocates, even as legislation and institutional support for their case accumulates exponentially"); Gupta and Mason (n 13), 446–447; Peter Cramton, David J. C. MacKay, Axel Ockenfels, and Steven Stoft (eds), *Global Carbon Pricing: The Path to Climate Cooperation* (Cambridge, MA: MIT Press, 2017); and Romain Weikmans, Harro van Asselt, and J. Timmons Roberts, 'Transparency Requirements under the Paris Agreement and Their (Un)Likely Impact on Strengthening the Ambition of Nationally Determined Contributions (NDCs)' (2020) 20(4) *Climate Policy* 511.

18 See, for example, Gupta and Mason (n 13), 450; and Esther Turnhout, Katja Neves, and Elisa de Lijster, '"Measurementality" in Biodiversity Governance: Knowledge, Transparency, and the Intergovernmental Science-Policy Platform on Biodiversity and Ecosystem Services (IPBES)' (2014) 46 *Environment and Planning A* 581, 583.

Requirements for state greenhouse gas MRV systems under international climate treaties

The UNFCCC, and especially the Kyoto Protocol, instigated the production of high-quality and independently reviewed information on the greenhouse gas emissions of Annex I parties to the UNFCCC (around three dozen developed countries). Even though the UNFCCC evidently intends to have all countries attain a high level of climate-related MRV, in fact only the Annex I parties have attained it in the course of the treaty's first quarter-century of operation. With the adoption of the Paris Agreement in 2015, developing countries are coming under pressure to quickly build permanent MRV systems of their own, of comparable quality, to be transparent about their greenhouse gas emissions.

MRV under the framework convention and Kyoto Protocol

Framework convention

Pursuant to Article 12 of the UNFCCC, all state parties have international reporting obligations. Reporting is to be made through National Communications and national greenhouse gas emission inventories submitted to the UNFCCC's Secretariat.[19]

Despite Article 12's all-inclusive language, for most of the UNFCCC's history, only Annex I parties have been obliged by COP decisions to meet a timetable for the submission of national emission reports. In the case of other countries (non-Annex I parties), the MRV obligation has remained subject to the provision of international finance to the developing countries needing it (which is almost all of them) and subject to soft deadlines; moreover, in the case of Least-Developed Countries (LDCs), national reporting has been at their discretion, an allowance that recognizes their severe capacity constraints.[20] It was not until after the decisions taken at the Copenhagen COP, in 2009, that the more advanced among the non-Annex I parties, including China, came to be included in an international reporting timetable with fixed deadlines, starting in 2014.[21]

For reasons soon to be made clear, we begin this section with a consideration of the guidelines for the preparation of National Communications by Annex I parties, which were first adopted in 1999.[22] A National Communication from an Annex

19 UNFCCC, Article 12(1).

20 Ibid., Article 12(5).

21 National Communications of non-Annex I parties are to be submitted every four years, and Biennial Update Reports every two years – as further explained below.

22 COP (UNFCCC), *Decision 4/CP.5, Guidelines for the Preparation of National Communications by Parties Included in Annex I to the Convention, Part II: UNFCCC Reporting Guidelines on National Communications*, FCCC/CP/1999/6/Add.1 (1999), para. 1. These guidelines are found in COP (UNFCCC), *Guidelines on Reporting and Review*, FCCC/CP/1999/7 (1999). Submission dates for Annex I National Communications are determined by decisions of the Conference of the Parties to the UNFCCC; see, e.g., COP (UNFCCC), *Decision 10/CP.13, Compilation and Synthesis of Fourth National Communications*, FCCC/CP/2007/6/Add.1 (2007).

I party, due roughly every four years, is to provide information on the state's policies and measures in furtherance of various UNFCCC requirements, consisting of mitigation, adaptation, financial assistance, and capacity-building support, as well as assorted other actions, such as actions undertaken domestically by the country to raise public awareness about climate change. An Annex I National Communication is also to provide information on a country's greenhouse gas emissions and its projected future emissions under different scenarios.[23] Preparing such a report every four years is no simple task. It is a whole-of-government effort. As will be discussed later, National Communications from non-Annex I parties follow a different set of requirements.

Conceptually distinct from the National Communication is the national greenhouse gas inventory. A national inventory is a highly technical report that reaches well beyond the "information" on greenhouse gas emissions required for National Communications. Article 4(1) of the UNFCCC calls on, we emphasize, *all states* to "Develop, periodically update, [and] publish . . . national inventories of anthropogenic emissions by sources and removals by sinks of all greenhouse gases . . . using comparable methodologies". These inventories, which Annex I parties must submit separately from their National Communications, and must do so annually,[24] detail activity data (emission sources reported according to standardized source types and quantified to show the amount of "activity" under each source), emission factors (the emissions per unit of activity per source), and the methods used by the state's agency charged with the preparation of the national inventory to collect, synthesize, and quality-control the required information.[25]

As we have indicated, despite the all-inclusive language of Article 4(1), Annex I parties took the lead in the preparation of national inventories. Preparing a state inventory of greenhouse gas emissions annually is an even heavier burden than preparing a national communication every few years. It implies the need for standing institutions and procedures, and it even imples legal regulation to ensure compliance.

Several decisions[26] and guidance documents[27] have elaborated the obligations of UNFCCC parties to report their greenhouse gas emissions. All countries'

23 COP (UNFCCC), *Guidelines on Reporting and Review*, section II, paras. 10–11.

24 COP (UNFCCC), *Decision 11/CP.4, National Communications from Parties Included in Annex I to the Convention*, FCCC/CP/1998/16/Add.1 (1998), para. 2.b; and COP (UNFCCC), *Decision 18/CP.8, Guidelines for the Preparation of National Communications by Parties Included in Annex I to the Convention, Part I: UNFCCC Reporting Guidelines on Annual Inventories*, FCCC/CP/2002/7/Add.2 (2002), para. 2.

25 See the sources cited below.

26 UNFCCC Subsidiary Body for Scientific and Technical Advice, *Guidelines for the Preparation of National Communications by Parties Included in Annex I to the Convention, Part I: UNFCCC Reporting Guidelines on Annual Inventories (Following Incorporation of the Provisions of Decision 13/CP.9)*, FCCC/SBSTA/2004/8 (2004); UNFCCC Subsidiary Body for Scientific and Technical Advice, *Updated UNFCCC Reporting Guidelines on Annual Inventories Following Incorporation of the Provisions of Decision 14/CP.11*, FCCC/SBSTA/2006/9 (2006); and UNFCCC, *Decision 24/CP.19, Revision of the UNFCCC Reporting Guidelines on Annual Inventories for Parties Included in Annex I to the Convention*, FCCC/CP/2013/10/Add.3 (2013).

27 IPCC, *Revised 1996 IPCC Guidelines for National Greenhouse Gas Inventories* (IPCC, 1996); IPCC, *Good Practice Guidance and Uncertainty Management in National Greenhouse Gas*

emission inventories "should" be *transparent* (transparency, in this context, means that assumptions and methods used to produce the inventory should be explained), *consistent* (including consistent with the state's inventories from past years), *comparable* with the inventories of other countries, *complete* (i.e., all sources and sinks and all gases should be covered), and *accurate* (with the uncertainties reduced as far as practicable).[28] We may refer to these as the UNFCCC's five MRV "principles". They are legally weak principles in that they are qualified by a "should", not a "shall".

A UNFCCC party must also highlight in its greenhouse gas report any emission source that was not considered in its inventory, despite its being listed in the applicable guidance documents issued by the Intergovernmental Panel on Climate Change (IPCC),[29] and explain why it was not considered.[30]

We now turn to the different set of guidelines for National Communications that apply to non-Annex I parties to the UNFCCC, and therefore to China. In applying these guidelines, non-Annex I parties "should take into account their development priorities, objectives and national circumstances".[31] They may attempt, if they wish, to meet the higher standard of reporting that applies to Annex I parties.[32] The essence of the rule, however, is that a non-Annex I party is to report at a standard that is reasonably feasible for it. The non-Annex I guidelines repeat the text of Article 4(7) of the UNFCCC, to the effect that the extent to which a non-Annex I party implements its commitment will depend on the implementation by Annex I parties of their commitment to provide financial support to those non-Annex I parties that need it to meet their MRV obligations.[33] The amount of climate finance available to developing countries for this purpose has increased over the years,[34] but developing countries have consistently complained that the amount of available climate finance is insufficient to cover all of their UNFCCC-related implementation expenses – leading to inaction, excused by Article 4(7).

A non-Annex I party "shall" include information on its greenhouse gas emissions in its National Communication.[35] It is not obliged to communicate a separate

---

Inventories (IPCC, 2000); IPCC, *Good Practice Guidance for Land Use, Land-Use Change and Forestry* (IPCC, 2003); and IPCC, *IPCC Guidelines for National Greenhouse Gas Inventories* (IPCC, 2006) (combining LULUCF and agriculture into a single Agriculture, Forestry, and Other Land-Uses (AFOLU) sector). Most of these manuals are technical in nature, extensive in scope, and complex in content, reflecting the complexity of the relevant issues and processes.

28 COP (UNFCCC), *Guidelines on Reporting and Review*, FCCC/CP/2002/8 (2002), Reporting Guidelines, para. 2.

29 See references below.

30 COP (UNFCCC), *Guidelines on Reporting and Review* (n 28), Reporting Guidelines, para. 28.

31 COP (UNFCCC), *Decision 17/CP.8, Guidelines for the Preparation of National Communications from Parties Not Included in Annex I to the Convention*, FCCC/CP/2002/7/Add.2 (2003), para. 1(b).

32 Ibid., para. 2.

33 Ibid., Annex, para. 27.

34 Global Environment Facility, 'Annual Performance Report 2019' <www.thegef.org/sites/default/files/council-meeting-documents/EN_GEF.ME_C56_Inf.01_Annual_Performance_Report_May_2019_0.pdf> accessed 3 December 2019.

35 COP (UNFCCC), *Decision 17/CP.8* (n 31), Annex, para. 2.

national inventory as Annex I parties are required to do. The greenhouse gas inventory of a non-Annex I party is thus updated only when the party prepares a new National Communication,[36] with the result that greenhouse gas emissions from non-Annex I parties have been reported relatively infrequently (see Table 7.1 for an example). Also, they do not meet the high standard of Annex I inventories, which further contributes to the patchiness of their record.

Non-Annex I parties were required to provide a greenhouse gas inventory for the year 1994 as part of their first National Communication. Their second National Communication was to include an inventory for the year 2000. LDCs were permitted to report their greenhouse gas emissions for any year that they could complete an inventory for, at their discretion.[37] We mention these facts to indicate just how far non-Annex I parties still have to go to build their national MRV systems. The hard separation between Annex I and non-Annex I parties has meant that some countries that could have made greater progress since 1992 were given no incentive to do so. China is surely in this category.

UNFCCC decisions require Annex I and non-Annex I parties to follow the IPCC's *Revised 1996 Guidelines for National Greenhouse Gas Inventories* and related guidance documents in preparing their inventories.[38] Parties may also use "national methods" that they consider better reflect their national circumstances, provided that such alternatives are compatible with IPCC guidance and are documented and based on science.[39] Non-Annex I parties are given greater flexibility in the application of the IPCC guidelines.[40] These guidelines are periodically revised, creating discontinuity and the need to recalculate past inventories, placing a further burden on developing countries. As noted below, the Paris Agreement requires states to use the IPCC's revised *2006 Guidelines*.

Kyoto Protocol

The Kyoto Protocol to the UNFCCC is, in all matters, narrower in scope than the Framework Convention. The Kyoto Protocol is directed primarily at Annex I parties. Article 7 of the Kyoto Protocol requires them to submit National Communications and greenhouse gas inventories. Since these two types of report are already submitted by Annex I parties pursuant to the UNFCCC, the Kyoto Protocol's requirements can be satisfied through the addition of annexes that cover the

36 In 2011 the UNFCCC's COP decided that the IPCC *Revised 1996 Guidelines*, in conjunction with the IPCC *Good Practice Guidance* and the IPCC *Good Practice Guidance for LULUCF*, should be used by developing countries for estimating and reporting anthropogenic emissions and removals (COP (UNFCCC), *Decision 4/CP.15, Methodological Guidance for Activities Relating to Reducing Emissions from Deforestation and Forest Degradation*, FCCC/CP/2009/11/Add.1 (2009)).

37 COP (UNFCCC), *Decision 17/CP.8* (n 31), Annex, para. 7.

38 COP (UNFCCC), *Guidelines on Reporting and Review* (n 28), Reporting Guidelines, para. 9.

39 Ibid., para. 10.

40 COP (UNFCCC), *Decision 17/CP.8* (n 31), Annex, paras. 8–9. The IPCC guidelines offer a default methodology that includes default emission factors and in some cases default activity data.

additional information required by the Kyoto Protocol. The Kyoto Protocol has had a significant impact on the MRV practices of Annex I parties but almost no impact on reporting by non-Annex I parties.[41] In other words, the latter countries were left to fall even further behind.

The Kyoto Protocol's main innovation in the area of MRV is the introduction of the concept of a "national system" for MRV. Each Annex I party is obliged by Article 5(1) of the Protocol to set up a national system for the preparation of its greenhouse gas inventory.[42] The concept of a national system for MRV is important because, under the Paris Agreement, all developing countries are expected to follow suit and develop a national system.[43] Thus, a technical term used by the Kyoto Protocol, which started off as an obligation of Annex I parties only, has evolved into a term of general application in the climate change regime.

A National Communication prepared pursuant to the Kyoto Protocol's guidelines must provide an account of the functioning of the state's national system, including a description of the institutional, legal, and procedural arrangements made to prepare the country's greenhouse gas inventory, such as the state's processes for collecting activity data, selecting emission factors, and ensuring the quality of the inventory.[44] Developing countries that are coming around to building national systems can of course learn from these accounts. They do not have to reinvent the wheel – and this applies equally to China.

Verification procedures

The UNFCCC established an international review system for state reports, including the greenhouse gas inventories. *Review* is considered to correspond to the third element of MRV. Nonetheless, the concept of review is weaker than verification. The latter suggests a procedure to determine *truth* in reporting. The international review procedure, by contrast, does not examine whether the domestic data provided by states is correct; instead, it takes them at face value. International review is therefore limited to verifying that the UNFCCC's reporting rules have been followed by the state, not that the reported information is true.

For most of the UNFCCC's history, international review was limited to the reports of Annex I parties. The National Communications of those parties have been subject to "in-depth review",[45] whereas Annex I national inventories undergo

41 The exception is that developing countries began to make reference in their reports to the Clean Development Mechanism, which was established by the Kyoto Protocol.

42 CMP (Kyoto Protocol), *Decision 19/CMP.1, Guidelines for National Systems under Article 5, Paragraph 1, of the Kyoto Protocol*, FCCC/KP/CMP/2005/8/Add.3 (30 November 2005).

43 See below.

44 CMP (Kyoto Protocol), *Decision 15/CMP.1, Guidelines for the Preparation of the Information Required under Article 7 of the Kyoto Protocol*, FCCC/KP/CMP/2005/8/Add.2 (30 November 2005), Annex, para. 30.

45 COP (UNFCCC), *Decision 2/CP.1, Review of First Communications from the Parties Included in Annex I to the Convention*, FCCC/CP/1995/7/Add.1 (1995), para. 2(a).

a process of "technical review".[46] The Kyoto Protocol added procedures for a so-called periodic review of National Communications and an "annual review" of inventories, which differ in small ways from the UNFCCC's procedures.[47] All these reviews are carried out by Expert Review Teams (ERTs), which are made up of (notionally) independent experts.[48] They are guided in their assessment by the UNFCCC's five MRV principles, discussed above.

In the past decade, in the wake of the Copenhagen Accord,[49] the UNFCCC's original MRV system has undergone two changes that increase the level of the MRV regulation of developing countries. First, a 2010 UNFCCC decision on National Communications from non-Annex I parties scheduled for submission in the 2013–2020 period requires submission at four-year intervals – the same frequency as developed countries.[50] Second, progress made with the mitigation pledges of UNFCCC parties, both developed and developing, for the 2013–2020 period is to be reported on,[51] and these reports are to undergo a degree of scrutiny through the processes of International Assessment and Review (for Annex I parties) and International Consultation and Analysis (for non-Annex I parties).[52] International Consultation and Analysis (ICA) represents the first time that non-Annex I parties were affected, however mildly, by the "V" in MRV.

Under the amended UNFCCC system, the reports on the mitigation progress by non-Annex I parties are called Biennial Update Reports (BURs). They are important in principle from an MRV perspective because their purpose is to link the mitigation policies of developing countries with observed impacts on those countries' greenhouse gas emissions. ICA for the BURs commenced in 2015. It consists of a technical analysis of a BUR, followed by a "facilitative sharing of views" among state parties in a forum convened by the UNFCCC's Subsidiary Body on Implementation (SBI).[53] Technical analysis is carried out by Teams of

46 COP (UNFCCC), *Decision 19/CP.8, Guidelines for the Technical Review of Greenhouse Gas Inventories from Parties Included in Annex I to the Convention*, FCCC/CP/2002/7/Add.2 (2002), paras. 1–2. The annex to the decision containing the guidelines is in fact a separate document: COP (UNFCCC), *Guidelines on Reporting and Review*.

47 CMP (Kyoto Protocol), *Decision 22/CMP.1, Guidelines for Review under Article 8 of the Kyoto Protocol*, FCCC/KP/CMP/2005/8/Add.3 (2005), Annex, para. 1.

48 COP (UNFCCC), *Decision 2/CP.1* (n 45), Annex II. For periodic reviews of National Communications under the Kyoto Protocol, see CMP (Kyoto Protocol), *Decision 22/CMP.1* (n 47), Annex, paras. 131–40; for annual reviews of greenhouse gas inventories, see ibid., Annex, paras. 50–117. On the independence of the ERTs, see Alexander Zahar, *International Climate Change Law and State Compliance* (Routledge, 2015).

49 COP (UNFCCC), *Decision 2/CP.15, Copenhagen Accord*, FCCC/CP/2009/11/Add.1 (2009).

50 COP (UNFCCC), *Decision 1/CP.16, The Cancun Agreements: Outcome of the Work of the Ad Hoc Working Group on Long-Term Cooperative Action under the Convention*, FCCC/CP/2010/7/Add.1 (2010), para. 60(b). Compliance is subject to "the prompt provision of financial resources to cover the agreed full costs" of preparing National Communications.

51 See ibid., para. 36.

52 Ibid., paras. 40(a) and 60(c). These are "should" rather than "shall" provisions.

53 COP (UNFCCC), *Decision 2/CP.17, Outcome of the Work of the Ad Hoc Working Group on Long-Term Cooperative Action under the Convention*, FCCC/CP/2011/9/Add.1 (2011), Annex IV, para. 3.

Technical Experts (TTEs).[54] As distinct from ERTs, the majority of TTE members are from developing countries, although they are still notionally independent experts.[55] The technical analysis aims to "increase transparency of mitigation actions and their effects" and identify capacity-building needs in order to improve non-Annex I reporting.[56] The costs of ICA are met from "additional" financial contributions by Annex I parties.[57] The effect of ICA is to increase pressure on developing countries to improve their MRV systems. The prospect of ICA may be assumed to have made China and other countries redouble their MRV efforts to plug gaps and avoid censure by other state parties at the SBI forum.

In summary, the establishment by the governments of Annex I parties of national systems for MRV, designed to serve the need for full and highly accurate greenhouse gas reporting under the Kyoto Protocol, facilitated within this relatively small group of countries a systematic increase in MRV quality. By contrast, because of the low expectations placed by the UNFCCC (and even more so by the Kyoto Protocol) on both the mitigation obligations and the MRV activities of non-Annex I parties, and notwithstanding the belated institution of the BUR/ICA process, the production of reliable information on greenhouse gas emissions at regular intervals by developing countries is still a rarity. The 2015 Paris Agreement aims to change this situation, as explained next.

"Enhanced" transparency under the Paris Agreement

The Paris Agreement sets a 2°C limit on global warming[58] and requires all parties to articulate their "contributions" to the mitigation of the causes of climate change, aiming to stay within the 2°C limit.[59] In addition, all parties are to submit information on a broad range of climate-related matters to the Conference of the Parties to the Paris Agreement (CMA) via the treaty's Secretariat, at fixed two-year intervals.[60] This process of information generation, submission, and review is called the Enhanced Transparency Framework (ETF). It is detailed in Article 13 of the Paris Agreement and in subordinate rules (discussed below).

As its name suggests, the ETF is expected to be an improvement on the UNFCCC's MRV system described in earlier sections. The ETF's purpose is to "provide a clear understanding of climate change action in the light of the objective of the Convention as set out in its Article 2, including clarity and tracking of progress

54 COP (UNFCCC), *Decision 20/CP.19, Composition, Modalities and Procedures of the Team of Technical Experts under International Consultation and Analysis*, FCCC/CP/2013/10/Add.2 (2013), Annex, para. 3.
55 Ibid., Annex, para. 5. In ERTs, developed and developing countries are equally represented.
56 Ibid., Annex, para. 15.
57 COP (UNFCCC), *Decision 2/CP.17* (n 53), paras. 58–59; and COP (UNFCCC), *Decision 20/CP.19* (n 54), para. 6.
58 Paris Agreement, Article 2(1)(a).
59 Ibid., Article 4(2).
60 Ibid., Article 13.

towards achieving Parties' individual nationally determined contributions".[61] The assumption is that the hoped-for clarity will increase trust among states and therefore mitigation ambition.

There are two main communication instruments under the Paris Agreement which all state parties must produce and are relevant to the MRV of state greenhouse gas emissions. The first is the Biennial Transparency Report (BTR).[62] The BTR is to include a national greenhouse gas inventory report, which may also be published as a self-standing report.[63] BTRs are due for submission starting at the end of 2024. Following the initial submission, a state must submit a BTR biennially.[64]

The second communication instrument has the character of a formal declaration: the Nationally Determined Contribution (NDC).[65] The NDCs, which were submitted in their initial form just before the Paris COP in 2015, are to go into effect in 2020. They are to be updated every five years thereafter.[66] The rules on NDC "accountability" (i.e. MRV of the specific promised "contributions" listed in the NDC) apply from the second round of NDCs onwards – that is, the period 2025–2030 and following.[67] The submitted information is to be reviewed by Technical Expert Review teams (TER teams).[68] The TER teams' first review reports[69] are due to appear in late 2026.[70]

Because this chapter is concerned with MRV systems for greenhouse gas emissions, the focus will be on the component of the BTR that constitutes the national greenhouse gas inventory report. Nevertheless, a brief account will be given in the next paragraph of MRV relating to the components of NDCs, as a developing country's national system must be set up to deal with this at the same time as it is building its system for tracking emissions.

NDCs are subject to two informational requirements: according to Article 4(8) of the Paris Agreement, they must be submitted with "the information necessary for clarity, transparency and understanding"; and, according to Article 4(13), they must be "accounted for" in their implementation.[71] The MRV of greenhouse gas

61 Ibid., Article 13(5).
62 COP (UNFCCC), *Decision 1/CP.21, Adoption of the Paris Agreement*, FCCC/CP/2015/10/Add.1, para. 90.
63 Paris Agreement, Article 13(7)(a); and CMA (Paris Agreement), *Decision 18/CMA.1, Modalities, procedures and guidelines for the transparency framework for action and support referred to in Article 13 of the Paris Agreement*, FCCC/PA/CMA/2018/3/Add.2, Annex, paras. 10 and 38.
64 Ibid., para. 3.
65 Paris Agreement, Article 4.
66 Ibid., Article 4(9).
67 COP (UNFCCC), Decision 1/CP.21 (n 62), para. 32; CMA (Paris Agreement), *Decision 4/CMA.1, Further guidance in relation to the mitigation section of Decision 1/CP.21*, FCCC/PA/CMA/2018/3/Add.1, para. 7.
68 Paris Agreement, Article 13(11).
69 CMA (Paris Agreement), *Decision 18/CMA.1* (n 63), Annex, para. 187.
70 Normally, review reports appear about two years after the submission deadline for state reports.
71 The rules on the two requirements are set out in CMA (Paris Agreement), *Decision 4/CMA.1* (n 67), Annexes I and II, respectively.

emissions is associated primarily with the second of these NDC-related requirements. It is a "shall" requirement for all parties to the Paris Agreement. In particular, states in their BTRs are to "account" for the implementation of their NDCs through monitoring and reporting the effects of their implemented measures.[72] TER teams are to serve as verifiers of the reports.[73] Because an NDC, as a state's contribution to the global effort, can be quite narrowly focused, the MRV of state greenhouse gas emissions pursuant to the Paris Agreement is broader than the MRV of NDC-based commitments on the mitigation of greenhouse gases. To clarify this point through an example, an NDC of a developing country might be concerned solely with reducing the emissions of carbon dioxide from the country's energy sector; therefore, NDC accounting for that country will be limited to the MRV of carbon dioxide mitigation from its energy sector. By contrast, the country is required by the terms of the ETF to carry out MRV of all its greenhouse gas emissions, including those associated with the country's NDC commitments. It is a much more laborious process.

MRV under the Paris Agreement

The Paris Agreement's requirements for the MRV of state greenhouse gas emissions, to be reported through the BTR, may be divided into what we will refer to as principles and rules. The main principles are as follows. The Paris Agreement's transparency framework (the ETF) is to be implemented "in a facilitative, non-intrusive, non-punitive manner, respecting national sovereignty and avoiding placing undue burden on Parties".[74] Transparency, accuracy, completeness, consistency, and comparability "shall" be promoted – and so is environmental integrity.[75] Moreover, reporting and transparency "shall" be improved over time.[76]

This group of six principles in the Paris Agreement consists of the UNFCCC's five MRV principles (see above), plus environmental integrity – all of them qualified by a "shall", not only by a "should", as under the UNFCCC. The six principles are listed in both the ETF Decision[77] and the NDC Accounting Decision,[78] which is evidence of their importance as principles to be followed.

Flexibility is another important principle in the ETF. Flexibility is to be granted to those developing countries that deem themselves in need of it in light of their (relatively low) capacity level.[79] The need for flexibility "is to be self-determined"

72 CMA (Paris Agreement), *Decision 4/CMA.1* (n 67), para. 17; and ibid., Annex II, para. 1(a).

73 CMA (Paris Agreement), *Decision 18/CMA.1* (n 63), Annex, paras. 6, 8, 147–149.

74 Ibid., para. 3; reflecting Paris Agreement, Article 13(2–3).

75 CMA (Paris Agreement), *Decision 18/CMA.1* (n 63), para. 3; see also CMA (Paris Agreement), *Decision 4/CMA.1* (n 67), Annex II, para. 2(a).

76 CMA (Paris Agreement), *Decision 18/CMA.1* (n 63), para. 3.

77 Ibid.

78 CMA (Paris Agreement), *Decision 4/CMA.1* (n 67), para. 11.

79 CMA (Paris Agreement), *Decision 18/CMA.1* (n 63), Annex, para. 3; reflecting Paris Agreement, Article 13(2–3).

by the developing country concerned.[80] The developing country claiming a need for flexibility is to indicate the provision(s) of the Paris Agreement's rules for which flexibility is to be granted, detail its capacity constraints, and provide an estimated time frame for overcoming those constraints.[81]

In addition to flexibility, the "special circumstances" of the LDCs and small island developing states "shall" be recognized.[82] Moreover, "support shall be provided to developing country Parties for the implementation of Article 13 and for the building of [their] transparency-related capacity . . . on a continuous basis".[83]

We will now outline the Paris Agreement's main rules on greenhouse gas reporting. As noted already, each party to the Paris Agreement must provide a national inventory of its greenhouse gas emissions as part of its BTR.[84] Each party "should" implement and maintain national inventory arrangements (i.e. a national system for the MRV of greenhouse gas emissions), including institutional, legal, and procedural arrangements for the continued estimation, compilation, and timely reporting of its national inventory. Although this is only a "should" (not a "shall") requirement, from the moment the Paris Agreement is launched in 2020, developing countries will come under constant pressure to set up permanent national systems. National inventory arrangements are allowed to vary, of course, "depending on . . . national circumstances and preferences".[85] But they must still deliver the same high-quality results and be comparable with each other. Thus, the amount of acceptable variation is necessarily limited.

Use of the IPCC's *2006 Guidelines*[86] as the basis for reporting under the Paris Agreement is a "shall" requirement, and each party "should" make every effort to use the recommended method (tier level) for key greenhouse gas categories in accordance with the *Guidelines*.[87] Each party "may" use nationally appropriate methods to calculate emissions if they better reflect its national circumstances and are consistent with the *Guidelines*.[88] Each party is "encouraged" to use country-specific and regional emission factors and activity data "where available".[89] If a state is unable to adopt a higher-level (more-stringent) tier method for a particular key emission category due to a lack of resources, the state "may" use a tier-1 approach (the lowest level). However, it is to document why it decided to do so, and it is to prioritize for future improvement the key categories for which the "good practice" method set out in the *Guidelines* could not be used.[90]

80 CMA (Paris Agreement), *Decision 18/CMA.1* (n 63), Annex, para. 6.
81 Ibid., para. 6.
82 Ibid., para. 3.
83 CMA (Paris Agreement), *Decision 18/CMA.1* (n 63), para. 7.
84 Ibid., Annex, para. 10.
85 Ibid., para. 18.
86 IPCC, *IPCC 2006 Guidelines*.
87 CMA (Paris Agreement), *Decision 18/CMA.1* (n 63), Annex, para. 21.
88 Ibid., para. 22.
89 Ibid., para. 24.
90 Ibid., para. 23.

There are many more rules, but here we will mention only the most significant ones. Each party "shall" report estimates of greenhouse gas emissions for all categories, gases, and carbon pools considered in its national inventory throughout the reported period on a gas-by-gas basis in units of mass at the most disaggregated level.[91] Each party "shall" report seven greenhouse gases,[92] but those developing countries that need "flexibility" in this respect (in the legal sense described earlier) have the discretion to instead report on the three main greenhouse gases only (carbon dioxide, methane, and nitrous oxide).[93] A state is to "account" for all categories of emissions for which specific actions are envisioned in its NDC.[94] Each party "shall" report on the sectors of energy, industrial processes and product use, agriculture, LULUCF (land use, land-use change, and forestry), and waste.[95] Each party "shall" report a consistent annual time series starting with 1990; however, a developing country that needs flexibility has the discretion to instead report data covering, at a minimum, the reference year or period used in its NDC as well as a consistent annual time series from at least 2020 onwards.[96]

How do these principles and rules affect China? From one point of view, they create a heavy burden for the country; from another, they allow it to avail itself of a generous amount of "flexibility". It remains to be seen whether China will enter a request for flexibility in connection with the preparation of its first BTR, due in 2024. For the purposes of this chapter, the Paris Agreement's principles and rules, outlined here, can be used as a standard, by scholars and government officials alike, by which to assess the level of development of China's MRV system for greenhouse gas emissions. Yet any judgement on China's "compliance" will have to be relative, not absolute: there are many dozens of countries that have made much less progress than China has on MRV of greenhouse gas emissions.

China's mitigation targets and national reporting performance

This section and the next will give a concise account of China's MRV system for greenhouse gas emissions. We will first present some contextual and background information.

91 Ibid., para. 47.

92 I.e. carbon dioxide, methane, nitrous oxide, hydrofluorocarbons, perfluorocarbons, sulphur hexafluoride, and nitrogen trifluoride.

93 CMA (Paris Agreement), *Decision 18/CMA.1* (n 63), Annex, para. 48.

94 CMA (Paris Agreement), *Decision 4/CMA.1* (n 67), Annex II, para. 3(a). The burden of this rule should be understood in the context of the rule that "Parties strive to include all categories of anthropogenic emissions and removals in their nationally determined contribution, and, once a source, sink or activity is included, continue to include it": ibid., para. 3(b).

95 CMA (Paris Agreement), *Decision 18/CMA.1* (n 63), Annex, para. 50.

96 Ibid., para. 57.

China's mitigation targets for greenhouse gas emissions

China has adopted greenhouse gas emission (and removal) targets at both the international and the domestic levels. At this point in time, all of China's targets concern carbon dioxide only.

China's internationally declared targets for 2020 are as follows:

> [To] lower carbon dioxide emissions per unit of GDP by 40% to 45% from the 2005 level, increase the share of non-fossil fuels in primary energy consumption to about 15% and increase the forested area by 40 million hectares and the forest stock volume by 1.3 billion cubic metres compared to the 2005 levels.[97]

China's internationally declared targets for 2030 are again limited to carbon dioxide.[98] This later target for carbon dioxide is a peaking target for the gas:

> [China intends] To achieve the peaking of carbon dioxide emissions around 2030 and mak[e] best efforts to peak early; to lower carbon dioxide emissions per unit of GDP by 60% to 65% from the 2005 level; to increase the share of non-fossil fuels in primary energy consumption to around 20%; and to increase the forest stock volume by around 4.5 billion cubic meters on the 2005 level.[99]

For its domestic target, China has adopted an 18% reduction target for the carbon dioxide intensity of its economy during the period of the 13th Five-Year Plan (2016–2020). This target has been broken down into provincial-level targets.[100]

China's national reports on greenhouse gas emissions

China has reported its progress towards its internationally declared targets for 2020 in several publications,[101] including in its most recent national reports to the UNFCCC.[102] Table 7.1 provides details on each of China's national reports and the greenhouse gas inventories incorporated therein.

China's five national reports to the UNFCCC each contain one inventory of greenhouse gas emissions covering a single year, for a total of five inventories completed so far.103 China's national emissions, excluding those from the LULUCF

97 China, *China's Intended Nationally Determined Contribution* (Beijing, National Development and Reform Commission, 2015), 3.

98 Although China has no internationally declared quantified targets for the next two most abundant anthropogenic greenhouse gases, namely methane and nitrous oxide, the government has expressed an intention "to control" them: ibid., 9.

99 Ibid., 5.

100 Maosheng Duan and Li Zhou, 'Key Issues in Designing China's National Carbon Emissions Trading System' (2017) 6(2) *Economics of Energy and Environmental Policy* 55, 58.

101 For example, China, *INDC* (n 97), 3–4; and National Development and Reform Commission, *China's Policies and Actions for Addressing Climate Change (2017)* (Beijing, October 2017), 1.

102 For example, Government of China, *Second Biennial Update Report on Climate Change* (Beijing, State Council, 2018), 3.

103 Ibid., 15–26.

Table 7.1 China's five national reports to the UNFCCC

Greenhouse Gas Inventory Year	*Inventory*	*Emissions, Excluding LULUCF ($MtCO_2$ eq.)*	*Recalculation?*
1994	1st National Communication (2004)	4,057	Not recalculated
2005	2nd National Communication (2012)	8,015	Recalculated in NC3[104]
2010	3rd National Communication (2018)	10,544	N/A (submitted 2018)
2012	1st Biennial Update Report (2016)	11,896	Not recalculated
2014[105]	2nd Biennial Update Report (2018)	12,301	N/A (submitted 2018)

Note: China received international funding support from the Global Environment Facility for all five of its reports to the UNFCCC.[106]

sector, as calculated in the five inventories, are shown in the third column, in megatons of carbon dioxide equivalent. Only one recalculation of an inventory has been carried out to date, even though three should have been recalculated.

Scholarly critiques of China's MRV performance

There has been a long line of criticism of China's climate-related reporting. We will summarize some of it here. The criticisms in the literature are never directly acknowledged by any of China's five reports to the UNFCCC.

Generally, on data transparency in China, Hsu and colleagues have reported on their experience of trying to develop environmental indicators for China, at China's request, circa 2010. They found that information on China's environmental conditions was negatively affected by inconsistencies in available monitoring systems and technical capacity due to economic constraints but also due to "political factors that limit information flows and processes; and institutional and

104 NC3 = 3rd National Communication. The recalculation applied the methods used for the 2010 inventory retrospectively to the 2005 inventory: China, *Third National Communication on Climate Change* (Beijing, 2018) 53–54.

105 China, *Second BUR* (n 102), 9. This inventory covers six gases (carbon dioxide, methane, nitrous oxide, hydrofluorocarbons, perfluorocarbons, and sulphur hexafluoride) from energy, industrial processes, agriculture, LULUCF, and waste. It follows mainly the *Revised 1996 IPCC Guidelines for National Greenhouse Gas Inventories*. For activity data, China drew mostly on official statistics, whereas emission factors were for the most part based on country-specific parameters, according to the report.

106 See Pat Hogan et al., *Tracking Emissions and Mitigation Actions: Current Practice in China, Germany, Italy, and the United States* (San Francisco: Climate Policy Initiative, 2012), 10; China, *First Biennial Update Report on Climate Change* (Beijing, 2016), 76; and China, *Second BUR* (n 102), 1.

organizational constraints that hamper information collection, processing, and dissemination".[107] They give examples of government-imposed restrictions on data disclosure, such as on provincial soil quality. The relevant information was collected by government agencies, but the researchers were not permitted to see it.[108] Moreover, they were not allowed to independently evaluate available statistics by using alternative data sources or through access to underlying methodologies and raw data. The authorities declared inaccessible detailed information on data collection methods and monitoring infrastructure for most indicators.[109] For the indicators and data that they were not prevented from obtaining, measurement systems were inconsistent and uneven between environmental issues. Hsu and colleagues found that only measurement systems concerning industrial efficiency were produced on a regular basis and in accordance with established methodologies in a transparent, consistent, and verifiable manner.[110]

Writing in 2012 about the preparation of China's 1994 inventory (published in 2004), Falconer and colleagues report on several shortcomings in the country's MRV system, including that the system lacked transparent expert and public review of data and methods.[111] They discuss the difficulties that they encountered in finding any documentation on detailed data sources or methodologies used for the 1994 inventory, and they report that they could find little description of quality control or quality assurance procedures applied in the inventory preparation.[112] Falconer and colleagues found that although penalties had been put in place to prevent data falsification, their enforcement was lax in practice.[113] They characterize the 1994 inventory as "substantially uncertain", due to a weak statistical foundation, limited scope, and observations and sampling that scored low on representativeness.[114]

Several studies – like those above, now a few years old – have cast doubt on China's statistics on coal consumption and carbon dioxide emissions. A 2013 study found that China's statistics on coal consumption were inaccurate.[115] In 2015, Zhu Liu and colleagues reported that the data sets for China's carbon emissions were not always reliable. Official national emission totals were found to differ from self-reported provincial statistics by 20%.[116] Their recalculations showed that total energy consumption in China was 10% higher in 2000–2012 than the

107 Angel Hsu, Alex de Sherbinin, and Han Shi, 'Seeking Truth from Facts: The Challenge of Environmental Indicator Development in China' (2012) 3 *Environmental Development* 39, 40.

108 Ibid., 43.

109 Ibid., 44.

110 Ibid., 44–45.

111 Falconer et al. (n 5), ii.

112 Ibid., 8–9.

113 Ibid., 10.

114 Ibid., 6.

115 Kexi Pan et al., 'Estimation of Coal-Related CO2 Emissions: The Case of China' (2013) 24(7–8) *Energy and Environment* 1309.

116 Zhu Liu et al., 'Steps to China's Carbon Peak' (2015) 522 *Nature* 279, 280.

value reported in China's national statistics.[117] They explain the discrepancies as arising mainly from conflicting estimates of energy consumption and emission factors, the latter being uncertain because of the paucity of available data on actual measurements of emission factors in the mix of Chinese fuels.[118]

From around the same period, a study by Korsbakken and colleagues observes that, according to preliminary official statistics released by China in 2015, coal consumption dropped by 2.9% in 2014; however, the researchers found that China's energy statistics are frequently revised and often contain large anomalies, implying high uncertainty.[119] The study identifies three inconsistencies in China's coal-use data that indicate uncertainty in the absolute consumption levels: a gap between total provincial coal-consumption data and national data; a gap between reported and apparent coal consumption (estimated from supply-side data); and continued growth in coal-intensive industrial products in 2014, despite the reported reduction in coal use.[120] Korsbakken and colleagues make an observation that is also true, as we noted earlier, of China's national reports to the UNFCCC, namely that there is rarely any official acknowledgement or explanation given by China's government of the significant inaccuracies that are evident in its revisions of its data or in recalculations in the scholarly literature (which the Chinese government may be assumed to be aware of).[121]

More-recent investigations have not eased the level of concern. Writing in 2018, Xuelan Zeng and colleagues found that uncertainty regarding data quality on greenhouse gas emissions persists because those regions in China that were not involved in setting up the pilot emission-trading schemes had insufficient experience in monitoring and reporting related to greenhouse gas emissions.[122] The study's detailed analysis finds that despite a rich foundation in regulations and the professionalization of the management of data-quality by enterprises in China,[123] shortcomings remain because of differences in data availability and uncertainty in data quality, such as in data on the carbon content of fuel at the enterprise level.[124]

As a last example, Yuli Shan and colleagues, also writing in 2018, report that no annual, officially published data on carbon dioxide emissions exist for China,[125] and the country's carbon dioxide emission accounts have not been well

117 Zhu Liu et al., 'Reduced Carbon Emission Estimates from Fossil Fuel Combustion and Cement Production in China' (2015) 524 *Nature* 335, 335.

118 Ibid., 335. The same conclusions were reached in a later study: Yuli Shan et al., 'New Provincial CO2 Emission Inventories in China Based on Apparent Energy Consumption Data and Updated Emission Factors' (2016) 184 *Applied Energy* 742, 743.

119 Jan Ivar Korsbakken, Glen P. Peters, and Robbie M. Andrew, 'Uncertainties around Reductions in China's Coal Use and CO2 Emissions' (2016) 6 *Nature Climate Change* 687.

120 Ibid., 688.

121 Ibid.

122 Xuelan Zeng et al., 'Data-Related Challenges and Solutions in Building China's National Carbon Emissions Trading Scheme' (2018) 18(S1) *Climate Policy* S90, S99.

123 Ibid., S97.

124 Ibid., S95.

125 Yuli Shan et al., 'China CO2 Emission Accounts 1997–2015' (2018) 5 *Scientific Data* 1.

documented.[126] They also found that hardly any of China's emission inventories are structured according to fossil fuel types and industrial sectors for the country and its 30 provinces.[127] They conclude that "Due to the poor quality of China's fossil fuel data, the fossil fuel consumption data also have large uncertainties".[128]

All this uncertainty found in China's nascent national system risks, of course, that China will overestimate its mitigation achievements. It also risks that other countries will not regard China's MRV system as reliable,[129] thus undermining (according to the transparency theory discussed earlier) any momentum towards greater global collective mitigation ambition.

China's MRV system for greenhouse gases

Is China's system ad hoc or legally regulated? How far is it likely to comply with the six MRV principles discussed earlier? Is there evidence of coordinated, steady improvement? These are some of the questions that we attempt to answer in this section.

Brief note on the hierarchy of laws and regulations in China

For ease of reference, the sources of law in China will be briefly recalled. They are listed here in order of importance:

> (1) the PRC Constitution; (2) "basic" laws (基本法) enacted by the National People's Congress; (3) other laws (法律) issued by the NPC Standing Committee; (4) NPC Standing Committee interpretations of the Constitution and basic law (立法解释); (5) State Council regulations and other documents having the force of law (行政法规); (6) ministerial regulations, standards, and rules (部门规章); (7) interpretations of the Supreme People's Court and Supreme People's Procuratorate; (8) regulations issued by the people's congresses and their standing committees of sub-national levels of government (地方法规); (9) sub-national level executive regulations and legal documents [including those of ministries and commissions under the State Council, and local governments at all levels[130]] (地方政府规章); (10) individual cases decided by Chinese courts; and (11) international agreements.[131]

126 Ibid., 2.

127 Ibid.

128 Ibid., 10–11.

129 See Heran Zheng et al., 'How Modifications of China's Energy Data Affect Carbon Mitigation Targets' (2018) 116 *Energy Policy* 337; and Kelly Sims Gallagher et al., 'Assessing the Policy Gaps for Achieving China's Climate Targets in the Paris Agreement' (2019) 10 *Nature Communications* 1, 2.

130 Tang et al. (n 14), S108.

131 Alex Wang, 'On the Evolution of Chinese Environmental Law and Governance' (2019) *Unpublished Conference Paper* available at SSRN <https://ssrn.com/abstract=3469184>, 4.

Even though "international agreements" is last in the list, it is obviously an important source of law in the context of MRV of national greenhouse gas emissions, given that most of the work that China has done in this area is in response to its international obligations.

Overview of China's MRV-related policy/legal instruments

China's regulatory framework for the MRV of greenhouse gas emissions consists, as shown in Table 7.2, of a constellation of policies and regulatory instruments. Almost all of them were promulgated during the period of the 12th and 13th Five-Year Plans (2011–2015 and 2016–2020, respectively). These MRV instruments aim to lay down three mechanisms to support national MRV of greenhouse gases: the development of the basic greenhouse gas inventory; greenhouse gas accounting for industrial enterprises; and an accountability system for controlling carbon dioxide emissions.

Table 7.2 List of MRV-related policy documents, methodologies, and standards in China[132]

Date of issue	*Document title*	*Issuing agency/Doc. no.*
Sep. 2010	Notice on Issues Related to Initiating and Preparing the Provincial GHG Emission Inventory (国家发改委办公厅关于启动省级温室气体排放清单编制工作有关事项的通知)	NDRC (2010)2350
Mar. 2011	Notice on Releasing the Guidelines for Preparation of Provincial GHG Inventory (Trial) (关于印发省级温室气体清单编制指南（试行）的通知)	NDRC (2011)1041
Jun. 2012	Interim Measures for Managing Voluntary GHG Emission Reduction Trading (温室气体自愿减排交易管理暂行办法)	NDRC (2012)668
May 2013	Opinions on Strengthening Statistical Work Related to Combating Climate Change (关于加强应对气候变化统计工作的意见)	NDRC and NBS (2013)937
Oct. 2013	Notice on Issuing GHG Emission Accounting Methods and Reporting Guidelines for Enterprises of the First Batch (Ten Industries) (Trial) (关于印发首批10 个行业企业温室气体排放核算方法与报告指南试行的通知)	NDRC (2013)2526

(*Continued*)

132 Compiled by the authors on the basis of government policy documents. Acronyms: NDRC (National Development and Reform Commission); NBS (National Bureau of Statistics); AQSIQ (General Administration of Quality Supervision, Inspection and Quarantine); SAC (Standardization Administration of China); GB (China National Standards); CNCA (China National Certification and Accreditation Administration).

Table 7.2 (Continued)

Date of issue	*Document title*	*Issuing agency/Doc. no.*
Nov. 2013	Notice on Launching Climate Change Statistical Operations (关于开展应对气候变化统计工作的通知)	NDRC and NBS (2013)80
Jan. 2014	Work Scheme for Statistical Operations Related to Addressing Climate Change (应对气候变化统计工作方案)	NBS (2014)7
Jan. 2014	Notice on the Preparation of GHG Emission Reports by Key Enterprises (Institutions) (关于组织开展重点企（事）业单位温室气体排放报告工作的通知)	NDRC (2014)63
Aug. 2014	Measures of Accountability Assessment with Regard to Fulfilment of the Control Targets related to CO_2 Emissions per Unit of GDP (单位国内生产总值二氧化碳排放降低目标责任考核评估办法)	NDRC (2014)1828
Dec. 2014	Notice on Issuing GHG Emission Accounting Methods and Reporting Guidelines for Enterprises of the Second Batch (Four Industries) (Trial) (关于印发第二批4 个行业企业温室气体排放核算方法与报告指南（试行）的通知)	NDRC (2014)2920
Jan. 2015	Notice on Launching Provincial GHG Inventory Preparation for the Next Stage (关于开展下一阶段省级温室气体清单编制工作的通知)	NDRC (2015)202
Jul. 2015	Notice on Issuing GHG Emission Accounting Methods and Reporting Guidelines for Enterprises of the Third Batch (Ten Industries) (Trial) (关于印发第三批10 个行业企业温室气体排放核算方法与报告指南（试行）的通知)	NDRC (2015)1722
Nov. 2015	General Guideline on GHG Emissions Accounting and Reporting for Industrial Enterprises (工业企业温室气体排放核算和报告通则) (effective on 1 June 2016)	AQSIQ and SAC GB/T32150–2015
Nov. 2015	Requirements of Greenhouse Gas Emission Accounting and Reporting: Parts 1 to 10 (温室气体排放核算与报告要求：第1部分 至第10部分) (effective on 1 June 2016)	AQSIQ and SAC GB/T32151.1–10–2015
Jun. 2017	General Specification of GHG Emission Verification for Industrial Enterprises[133] (行业企业温室气体排放核查通用规范)	CNCA RB/T211–2016

133 Following the General Specification, the CNCA released 11 sector-based specifications for GHG emission verification in two batches. The first batch has been implemented since October 2018, and the sectors include iron and steel, petrochemicals, power grid, electricity generation, calcium carbide, synthetic ammonia, methanol, and ethylene. The second batch has been implemented since December 2018, and the sectors include plate glass, cement, and ceramics (as of 31 May 2019).

Government departments involved in China's national system for MRV

The government agencies tasked with the establishment of China's MRV system for greenhouse gas emissions include the National Development and Reform Commission (NDRC) and the National Bureau of Statistics (NBS).[134] Table 7.3 summarizes the current situation.

The NBS "organizes relevant departments to provide basic statistical data, coordinates relevant industry associations and typical enterprises to provide relevant data and establishes [a] national greenhouse gas inventory database to support the preparation of the National GHG Inventory database and data management".[135] As the system has developed and the accounting methodologies have matured and become officially recognized, other government agencies, including the AQSIQ (Administration of Quality Supervision, Inspection, and Quarantine), the SAC (Standardization Administration of China), and the CNCAA (China National Certification and Accreditation Administration) have stepped up to stipulate national standards related to the verification of greenhouse gas emissions from enterprises and institutionalize related provisions.[136]

In 2014, the NBS, along with the NDRC, Ministry of Science and Technology, and industry associations, set up a 23-member Leading Group on Climate Change Statistics.[137] Since 2016, full-time personnel in the statistics departments of 27 of China's provinces have been made responsible for statistical and accounting work related to climate change.[138] In 2018, responsibility for addressing climate change was reassigned from the NDRC to the newly established Ministry of Ecology and Environment.[139] China now claims to have established a basic national system for MRV pursuant to the UNFCCC.[140]

134 China, *Third NC* (n 104), 48.

135 Ibid., 27.

136 General Guideline on GHG Emissions Accounting and Reporting for Industrial Enterprises (工业企业温室气体排放核算和报告通则), AQSIQ and SAC GB/T32150–2015 (effective on 1 June 2016); Requirements on Greenhouse Gas Emission Accounting and Reporting: Parts 1 to 10 (温室气体排放核算与报告要求：第1部分 至 第10部分), AQSIQ and SAC GB/T32151.1–10–2015 (effective on 1 June 2016); General Specification on GHG Emissions Verification for Industrial Enterprises (行业企业温室气体排放核查通用规范), CNCA RB/T211–2016, June 2017.

137 China, *First BUR* (n 106), 102.

138 NDRC (n 101), 52.

139 China, *Second BUR* (n 102), 1.

140 China, *Third NC* (n 104), 27. See also China, *Second BUR* (n 102), 5 ("China has built relatively stable institutions for technical support and organized core teams of experts, which have provided institutional guarantees for the preparation and submission of national communications and biennial update reports").

Table 7.3 Main bodies in China involved in the preparation of the country's reports to the UNFCCC[141]

Overall responsibility for National Communications, Biennial Update Reports, and national GHG inventory	Ministry of Ecology and Environment (National Authority Addressing Climate Change)
GHG inventory for energy	NCSC, Energy Research Institute of National Development and Reform Commission (ERI), Fudan University, China Special Equipment Inspection and Research Institute (CSEI)
GHG inventory for industrial processes	Tsinghua University, Foreign Economic Cooperation Office, Ministry of Ecology and Environment
GHG inventory for agriculture (livestock)	Institute of Environment and Sustainable Development in Agriculture, Chinese Academy of Agricultural Sciences
GHG inventory for agriculture (croplands)	Institute of Atmospheric Physics, Chinese Academy of Sciences
GHG inventory for LULUCF	Institute of Forest Ecology, Environment and Protection, Chinese Academy of Forestry; Research, Planning and Design Institute of the National Forestry and Grassland Administration; Research Institute of Forestry New Technology, Chinese Academy of Forestry; Institute of Environment and Sustainable Development in Agriculture, Chinese Academy of Agricultural Sciences; Institute of Atmospheric Physics, Chinese Academy of Sciences
GHG inventory for waste	Chinese Research Academy of Environmental Sciences
National GHG inventory database	National Center for Climate Change Strategy and International Cooperation

Although two National Communications had been produced by 2012, China's national system for the MRV of greenhouse gas emissions, characterized by systematic and standardized practices, was not put into operation until 2014.[142]

Current design and functioning of China's national MRV system

China's central government initiated an effort to establish a greenhouse gas accounting system in 2010. This began with an NDRC *Notice* demanding that

141 Adapted from China, *Second BUR* (n 102), 8.

142 On the ad hoc arrangements used to produce the first inventory, of 1994, see Falconer et al. (n 5), 9 ("The inventory team designed over 1,000 questionnaires during the first inventory preparation to collect information on activity data and emission factors from the sectors where routine collection of relevant data were unavailable or the data reliability was poor").

the provincial authorities set up work plans and programmes to advance the development of provincial greenhouse gas inventories.[143] They were to draw on experiences obtained in seven provinces that produced inventories through experimentation.[144] The 2010 initiative was made without sufficient guidance from the central government in terms of greenhouse gas inventory and accounting methods, and the provincial authorities in general lacked the technical capacity to do the job themselves.[145]

In 2011, the nationwide MRV initiative for greenhouse gases was endorsed in the 12th Five-Year Plan. It required the relevant government agencies to set up a national accounting system for greenhouse gas emissions and to strengthen the associated statistical system.[146] The NDRC was assigned the task of developing China's greenhouse gas accounting system. The *Guidelines for Preparation of Provincial GHG Inventory (Trial)*, promulgated by the NDRC in 2011, sought to enhance the scientific standardization and implementation of the activities relating to the development of provincial greenhouse gas inventories and to provide guidance on the making of inventories with consistent and comparable results and having a transparent format.[147]

According to the *Guidelines*, six greenhouse gases were to be included for accounting purposes. The emission sources and carbon sinks determined at the provincial level encompassed mainly energy activities, industrial production, agricultural activities, land use change and forestry, and urban waste treatment.[148]

The effort towards the institutionalization of greenhouse gas accounting was also reflected in a joint announcement made by the NDRC and NBS in October 2013, by which were established an accounting index (a list of items subject to greenhouse gas accounting) and an allocation of responsibilities.[149] The accounting index incorporated five categories of data collection: climate-change-related impacts, adaptation, mitigation, finance, and management. Under each category, detailed indicators were specified to guide the collection of data, and each indicator was linked directly to one or more central government agencies that were to be responsible for collecting the data.[150]

143 NDRC, Notice on Initiating the Compilation of Provincial Greenhouse Gas Inventory (关于启动省级温室气体清单编制工作有关事项的通知), NDRC Order No. (2010)2350, September 2010.

144 The seven provinces that prepared the inventories for experimentation purposes were Guangdong, Hubei, Liaoning, Yunnan, Zhejiang, Shaanxi, and Tianjin.

145 China, *First BUR* (n 106).

146 State Council, 'The 12th Five-year Plan for Economic and Social Development of the People's Republic of China' (中华人名共和国国民经济和社会发展第十二个五年规划纲要), March 2011.

147 Notice on Releasing the Guidelines for Preparation of Provincial GHG Inventory (Trial) (关于印发省级温室气体清单编制指南（试行）的通知), NDRC Order No. (2011)1041, March 2011.

148 Ibid. The six types of GHGs are CO_2, CH_4, N_2O, HFCs, PFCs, and SF_6.

149 Opinions on Strengthening Statistical Work related to Combating Climate Change (关于加强应对气候变化统计工作的意见), NDRC and NBS Order No. (2013)937, October 2013.

150 Ibid., annex 1 and 2.

With the overarching design of the accounting system in place, the NDRC and NBS, in November 2013, requested the subordinate provincial authorities to carry out data collection on the basis of the accounting index and indicators through a bottom-up approach, meaning that data collection and reporting under each category was to be delegated to the corresponding government agencies at lower levels.[151] The NBS designed, and required the use of, a standardized format for reporting, aiming for the data collection to be conducted in a consistent and comparable manner.[152] The standardization process was intended to facilitate the discovery of loopholes and inconsistencies to enable improvements to be made as the implementation of the system progressed.[153]

The NDRC/NBS regulation of November 2013, along with the NBS's reporting scheme (January 2014), could be said to mark the official implementation of China's greenhouse gas accounting system. The new system aimed to institutionalize and routinize greenhouse gas accounting in a comprehensive approach.

Greenhouse gas emissions from enterprises were to be given the highest attention considering their high contribution to greenhouse gas emissions and also the need to further develop and improve MRV methods associated with various industrial sectors.[154] Because of the significant contribution of industrial production to greenhouse gas emissions in China and the general lack of technical expertise on accounting (including the lack of MRV methods), the NDRC promulgated a number of accounting methods applicable to different industrial sectors and initiated greenhouse gas accounting targeted at key enterprises and institutions.

The national greenhouse gas reporting scheme for key enterprises and institutions is based on the NDRC's 2014 *Notice on the Preparation of GHG Emission Reports by Key Enterprises (Institutions)*.[155] The *Notice* requires key enterprises and institutions (defined as enterprises or institutions with emissions exceeding 13,000 tons of carbon dioxide equivalent, or a total energy consumption exceeding 5,000 tons of standard coal equivalent, in the baseline year of 2010)[156] to report the greenhouse gas emissions for which they are annually responsible, according to the approved accounting methods and reporting guidelines for the specific industrial sector.[157]

151 Notice on Launching Climate Change Statistical Operations (关于开展应对气候变化统计工作的通知), NDRC and NBS Order No. (2013)80, November 2013.

152 Work Scheme for Statistical Operations Related to Addressing Climate Change (应对气候变化统计工作方案), NBS Order No. (2014)7, January 2014.

153 Ibid.

154 Opinions on Strengthening Statistical Work related to Combating Climate Change (n 149), part 4.

155 Notice on the Preparation of GHG Emission Reports by Key Enterprises (Institutions) (关于组织开展重点企（事）业单位温室气体排放报告工作的通知), NDRC Order No. (2014)63, January 2014.

156 Ibid., part 3.

157 See, e.g., Notice on Issuing GHG Emission Accounting Methods and Reporting Guidelines for Enterprises of the First Batch (Ten Industries Involved) (Trial) (关于印发首批10 个行业企业温室气体排放核算方法与报告指南试行的通知), NDRC Order No. (2013)2526, October 2013.

To date, the NDRC has published, on a trial basis, 23 greenhouse gas accounting methods applicable to various industrial sectors.[158]

The NDRC's *Notice* requires six types of greenhouse gas emissions to be reported,[159] as well as four sources of emissions: fuel combustion, production processes, purchased electricity and heat, and generated electricity and heat.[160] The reporting entities are also obliged to provide information on activity-level data and how it is sourced and on the emission factors and how they are determined.[161]

Based on the NDRC's *Notice*, the requirements and rules underpinning China's greenhouse gas accounting practices have been endorsed and instituted by the affected government agencies, including the AQSIQ, SAC, and CNCAA, which oversee quality assurance and standardization processes at the national level.[162] The binding nature of national standards on the operation of enterprises is likely to standardize their greenhouse gas accounting and reporting practices, thus making the overall results more reliable, consistent, and comparable.

As of 2016, the AQSIQ and the SAC have promulgated 12 sector-specific national standards for greenhouse gas emission accounting and reporting. The sectors are electricity generation, power grids, civil aviation, and the production of magnesium, aluminium, iron, steel, plate glass, cement, ceramics, petrochemicals, coal, and textiles and garments.[163] For sectors for which specific guidelines have not yet been established, the *General Guideline on GHG Emissions Accounting and Reporting for Industrial Enterprises* applies. The rules encourage activity-level data and emission factors that are sourced and determined by field measurement at each reporting installation.[164]

To motivate officials' compliance with the rules, as part of the assessment regime for meeting carbon dioxide mitigation targets, government officials were required to be evaluated for their performance.[165] In August 2014, the NDRC

158 Apart from the first batch of accounting methods, the NDRC has issued another two batches: Notice on Issuing GHG Emission Accounting Methods and Reporting Guidelines for Enterprises of the Second Batch (Four Industries Involved) (Trial) (关于印发第二批4 个行业企业温室气体排放核算方法与报告指南（试行）的通知), NDRC Order No. (2014)2920, December 2014; Notice on Issuing GHG Emission Accounting Methods and Reporting Guidelines for Enterprises of the Third Batch (Ten Industries Involved) (Trial) (关于印发第三批10 个行业企业温室气体排放核算方法与报告指南（试行）的通知), NDRC Order No. (2015)1722, July 2015. The third batch also includes a general guideline that applies to industries that are outside of the 23 sectors, with specific guidelines for reporting.

159 Ibid., part 4.

160 Ibid., part 4(2).

161 Ibid., part 4(3), para. 1.

162 General Guideline of the GHG Emissions Accounting and Reporting for Industrial Enterprises (n 136). Worth noting is that the General Guideline includes NF_3 in the list of GHGs that are subject to accounting and reporting.

163 Ibid.

164 Ibid.

165 Measures of Accountability Assessment with Regard to Fulfilment of the Control Targets related to CO2 Emissions per Unit of GDP (单位国内生产总值二氧化碳排放降低目标责任考核评估办法), NDRC Order No. (2014)1828, August 2014.

and NBS, along with the Ministry of Industry and Information Technology, the National Energy Administration, and the State Forestry Administration, formed an assessment and evaluation working group to conduct preliminary reviews of reports submitted by each province, municipality, or autonomous region, with onsite inspection as necessary.[166] The assessment results, published yearly, were required to be taken into account in the performance evaluation of local government officials.[167] Provincial governments with poor evaluation results are under an obligation to submit a written report within a month of the assessment to propose measure, to rectify the problem.[168] Localities or government officials that conceal or falsify data during data collection and reporting must be "named and shamed".[169] In cases of serious misconduct or dereliction of duty, those responsible must be referred to the relevant supervisory authority for further investigation.[170]

Quality assurance and third-party verification of greenhouse gas reporting in China

To ensure the quality of the reporting of key enterprises and institutions in accordance with the 2014 *Notice*, the CNCAA, in June 2017, adopted the *General Specification on GHG Emissions Verification for Industrial Enterprises*. They contain detailed provisions that regulate the practices of verifiers and oversee the verification process.

The verification process consists of four stages: preparation, planning, implementation, and reporting.[171] These components mirror the procedural requirements of verification under China's domestic emission-offset scheme, which is designed and operated in accordance with the Clean Development Mechanism under the Kyoto Protocol.[172] There is evidence that China's participation in the CDM has enabled China's domestic verifiers to assimilate technical expertise from their international counterparts and help key stakeholders build their capacity.[173] In particular, the experience has increased the regulatory capacity of the NDRC and CNCAA, which have been extensively involved in regulating verification activities and standardizing the accreditation of verification agencies.

The *General Specification* incorporates a range of detailed practical standards and reference examples to regulate and guide the conduct of verifiers.[174] We will

166 Ibid., art. 7.
167 Ibid., art. 8.
168 Ibid., art. 10.
169 Ibid., art. 11.
170 Ibid.
171 General Specification on GHG Emissions Verification for Industrial Enterprises (n 136), part 6.
172 Hao Zhang and Christopher Arup, 'Beyond the CDM: Regulating China's Domestic Offset Scheme' (2015) 45(1) *Environmental Law Reporter* 10049–10063.
173 Ibid. See also Xiaoyi Jiang, *Legal Issues for Implementing the Clean Development Mechanism in China* (Springer, 2013).
174 General Specification on GHG Emissions Verification for Industrial Enterprises (n 136), part 5.1 and 5.2.

mention some of them here. The verification agencies must in the first place be qualified. The core requirement for accreditation is technical expertise and having available the minimum number of qualified personnel needed to carry out verification activities.

The first stage of the verification process – preparation for verification – includes two elements: first, the establishment of a review team and, second, a preliminary review of the greenhouse gas accounting report, focusing on the suitability of the accounting method used and the report's compliance with the selected method. A verification contract is signed between an applicant (the key enterprise or institution) and the verification agency.[175] The second stage is planning, which essentially involves the development and design of a verification plan. Document review and field measurement are mandatory verification methods.[176] Determining the scope of sampling requires assessing the data quality, the validity of emission factors used, the complexity of the accounting methodology, and the uncertainty of the verification results.[177] The key components of the third stage – the implementation stage – are verifying the emission boundary, the emission sources, the method used, and the accuracy of the data. Essentially the verification centres on ensuring that the data, including activity-level data and emission factors, are reasonably sourced and accurate.[178] In the last stage, the final report is subject to an internal review by an independent technical reviewer to examine whether the verification was carried out on the basis of the standards and requirements set out in the *General Specification*.[179] To ensure the independence of the technical review, the reviewer is prevented from participating in the onsite verification or the preparation of the verification report.[180]

Conclusion

Because regular, peer-reviewed reporting was not a requirement for developing countries under the UNFCCC or the Kyoto Protocol, the Paris Agreement's requirement that all developing countries commence quality reporting presents a great challenge for them. Most are unprepared to do what the new regime requires of them, because they have not set up permanent, budgeted, institutions with a capacity to produce a BTR every two years.

For this reason, the Paris Agreement's implementation rules afford "flexibility" to developing countries. In fact, the impact of the rules is moderated for all parties through the principle that no party is to be placed under "undue burden" by the

175 Ibid.
176 Ibid., part 7.4.
177 Ibid., part 7.5(1).
178 Ibid., part 8.2.4. Activity level data include real-time monitoring data, statistical data, onsite measurement data, and laboratory test data.
179 Ibid., part 9.2.1.
180 Ibid., part 10.2.

ETF.[181] From China's point of view, the key point about the implementation rules is that for all the burden they seem to imply at first sight, most of the "shall" requirements are qualified by exceptions that benefit developing countries.[182]

Although China, legally speaking, is operating in a flexible international environment as far as the MRV of greenhouse gas emissions is concerned, it has nevertheless made significant improvements to its national MRV system since 2012.[183] The government has acknowledged that it will need to continue to improve its national system, including by strengthening its legal regulation.[184] China has even stated that "Domestic MRV capacity building is crucial to how effectively developing countries can address climate change".[185] Therefore, there is no want of awareness on the part of China.

In closing, we agree with Renhu Tang and colleagues that given the European Union's experience,[186] the importance of having a clear legal framework that provides a strong basis for enforcing MRV requirements on different stakeholders is beyond doubt. China does not have this at present. It is crucial for the country to create a general framework legislation "that is legally binding and of a sufficiently high legal status to facilitate legal enforcement of the MRV requirements and increase the compliance levels of MRV activities".[187] It is also important for implementing regulations to include requirements on monitoring and reporting of those regulations, with penalties for noncompliance, and with judicial overview, to ensure compliance with the MRV principles.[188]

181 CMA (Paris Agreement), *Decision 18/CMA.1* (n 63), Annex, part I.B.3(a).

182 E.g. ibid., paras. 11 (exception for LDCs and SIDS), 23 (higher-level tiers need not be used by those countries "lacking resources"), 29 (flexibility on uncertainty estimation for developing countries), 34 (flexibility on elaboration of QA/QC plan for developing countries), 48 (not all of the greenhouse gases in the standard set need be reported by developing countries), 57 (exception from consistent annual time series since 1990), 85 (exception from estimation of mitigation impacts of policies and measures), 92 (covering paras. 93–101: exception from reporting of emission projections), and 157 (LDCs and SIDS need not be individually reviewed but may be group-reviewed if they so prefer).

183 See, for example, China, *Second BUR* (n 102), 9, 10, 22, and 23.

184 China, *INDC* (n 97), 6.

185 China, *First BUR* (n 106), 97.

186 See Directive 2003/87/EC of the European Parliament and of the Council of 13 October 2003 establishing a scheme for greenhouse gas emission allowance trading within the European Community and amending Council Directive 96/61/EC, OJ L 275, 25 October 2003, 32–46.

187 Tang et al. (n 14), S106.

188 Ibid.

8 China's legal and policy framework for adaptation

Challenges and responses

Xiangbai He

Introduction

As a developing country with a large population, uneven economic development, and a fragile ecological environment, China is vulnerable to the impacts of climate change.[1] According to a report by the Hong Kong and Shanghai Banking Corporation (HSBC) titled *Scoring Climate Change Risk*, China is the third most vulnerable country among the G-20 countries (after India and Indonesia) in terms of its exposure, sensitivity, and adaptive capacity to climate change.[2] Some of the observed and projected impacts outlined in the Fifth Assessment Report of the Intergovernmental Panel on Climate Change (IPCC) are an exacerbation of water scarcity in Northern China, glaciers retreating on the Qinghai–Tibet plateau, sea-level rise in Shanghai, changes to crop productivity, increased flood risk, more-frequent and more-intense heat waves, and changes in the geographical distribution of vector-borne diseases.[3] China's Third National Communication on climate change states that during the past ten or more years, there has been a significant decrease in precipitation in Southwestern China.[4] An increase in high-temperature events, a decrease in low-temperature events, and an increase in the frequency and strength of extreme heavy-precipitation events are expected in most regions.[5] These threats are particularly severe in agriculture and animal husbandry, forestry, natural ecological systems and water resources, and coastal and ecologically fragile zones.[6]

1 China's National Climate Change Programme (中国应对气候变化国家方案), National Development and Reform Commission (NRDC) Order No. 17, June 2007.

2 Zoe Knight, Nick Robins, and Wai-Shin Chan, 'Scoping Climate Change Risks: Which Countries Are Most Vulnerable?' (HSBC Global Research Report, HSBC, 2011) 4.

3 IPCC, *Climate Change 2014: Impacts, Adaptation, and Vulnerability* (Contribution of Working Group II to the Fifth Assessment Report of the Intergovernmental Panel on Climate Change) 1333–1334.

4 Ministry of Ecology and Environment (MEE), The Third National Communication on Climate Change (中华人民共和国气候变化第三次国家信息通报), 2018.

5 Ibid.

6 Information Office of the State Council, China's Policies and Actions for Addressing Climate Change, 2018 (中国应对气候变化的政策与行动(2018)), October 2018.

Yet China's response to climate change has generally centred on mitigation, with adaptation marginalized by the government in relation to target setting, financing, and institutional frameworks.[7] This has contributed to a low capacity for adaptation to climate change in natural resource planning and a lack of mainstreaming of in the climate adaptation measures in natural resource management.

International treaties and policies on climate change in the 1990s and early 2000s were also preoccupied mainly with the mitigation of greenhouse gases (GHGs), and adaptation took second place.[8] The United Nations Framework Convention on Climate Change (UNFCCC) treats adaptation in the narrowest sense, underscoring the treaty's intention to focus on reducing the causes of climate change.[9] Adaptation in the UNFCCC was not considered a policy objective in itself but rather a threshold for determining the level and time frame of mitigation.[10] The response to climate change adaptation gained increasing attention after the 2001 Marrakesh Accords.[11] However, its development has not kept up with the interest in mitigation.[12]

The 2015 Paris Agreement, a cornerstone of global climate governance for the coming decades, sets an adaptation goal to enhance adaptive capacity and resilience and to reduce vulnerability, with a view to contributing to sustainable development.[13] It is a milestone in terms of the ongoing efforts to make adaptation an equal priority with mitigation by widening the normative framing around adaptation, integrating stronger adaptation commitments from states, being explicit about the multilevel nature of adaptation governance, and strengthening transparency mechanisms for assessing adaptation progress.[14] The Paris Agreement and the intensifying climatic impacts and low adaptive capacity have put pressure on the Chinese government to undertake adaptation measures.[15] This chapter examines

7 He Xiangbai, 'China's Legal Framework on Climate Change Adaptation: Ambitious or Conservative?' (突破还是保守 – 评我国适应气候变化的法律框架) (2015) 2 *Nanjing University Law Review* (南京大学法律评论) 230–244.

8 Jan McDonald, 'A Short History of Climate Adaptation Law in Australia' (2014) 4 *Climate Law* 150–167.

9 Lisa Schipper, 'Conceptual History of Adaptation in the UNFCCC Process' (2006) 15(1) *Review of European, Comparative and International Environmental Law* 82–92.

10 Article 2 of the UNFCCC defines the goal of the convention that such a level of mitigation should be achieved within a time frame sufficient to allow ecosystems to adapt naturally to climate change, to ensure that food production is not threatened, and to enable economic development to proceed in a sustainable manner.

11 Irene Suarez and Jennifer Huang, 'Addressing Adaptation in a 2015 Climate Agreement' (Center for Climate and Energy Solutions, 2015).

12 Farhana Yamin and Joanna Depledge, 'The International Climate Change Regime: A Guide to Rules, Institutions and Procedures' (Institute of Development Studies, University of Sussex, December 2003).

13 UNFCCC, Paris Agreement, FCCC/CP/2015/L.9,12 December 2015.

14 Alexandra Lesnikowski et al., 'What Does the Paris Agreement Mean for Adaptation?' (2017) 17(7) *Climate Policy* 825–831.

15 UNFCCC, Katowice Rulebook, FCCC/CP/2018/10/Add.1, 19 March 2019.

from a legal perspective what has been done by the Chinese government to support the goal of adaptation.

Understanding the challenges of adaptation in China's legal system

Differences between mitigation and adaptation

Different from traditional environmental pollution, the large-scale, complex, and uncertain nature of climate change brings great challenges to China's legal system. Furthermore, it opens up two areas – mitigation and adaptation – which differ in their causes and effects, sectors involved, spatial-temporal scales, goals and approaches, and assessment criteria.[16] These differences need to be considered by, and reflected within, the legal framework and institutional arrangements. Adaptation is more ambiguous than mitigation, as it is difficult to define, identify, and track precisely what constitutes adaptation.[17] It has no standard reference metrics (as mitigation does); therefore, metrics to determine adaptation needs and the effectiveness of adaptation can be disputed. Compared with mitigation, which often relies on a centralized approach to emission reduction, adaptation follows a decentralized, often-localized process to identify adaptation needs and priorities that are sensitive to context.[18] Adaptation shares mitigation's long-term planning horizons, but it involves different levels of government, governmental branches, and non-state actors.[19]

With a strong focus on energy security and economic development, China's policy and law have placed considerable effort on energy-production reforms, including GHG-emission-intensity reduction, renewable-energy development, low-carbon-economy development,[20] which has led to a mitigation-oriented legal and institutional system in China. Prominent examples are the Energy Conservation Law, the Renewable Energy Law, and various specialized medium- and long-term plans for energy development and conservation. Yet adapting to climate change impacts that are already "locked in" is also crucial, especially for vulnerable sectors such as water resources and agriculture.[21] As Pielke et al. have said, "new ways of thinking about, talking about and acting on climate change are

16 Rob Swart and Frank Raes, 'Making Integration of Adaptation and Mitigation Work: Mainstreaming into Sustainable Development Policies?' (2007) 7(4) *Climate Policy* 288, 291.

17 Lesnikowski et al. (n 14).

18 Ibid.

19 Julie Teebken, 'Ten Years, Nine Floods: Local-Level Climate Adaptation in China', *News Security Beat* (29 March 2019) <www.newsecuritybeat.org/2018/03/ten-years-floods-local-level-climate-adaptation-china/> accessed 29 October 2019.

20 Scott Moore, 'Strategic Imperative? Reading China's Climate Policy in terms of Core Interests' (2011) 23(2) *Global Change, Peace and Security* 147.

21 Organization for Economic Cooperation and Development, 'Integrating Climate Change Adaptation into Development Co-operation: Policy Guidance' (2009) 48.

necessary if a changing society is to adapt to a changing climate".[22] Nonetheless, adaptation does not necessarily require a parallel legal system separate from the mitigation-dominated climate change framework.

Legal challenges presented by adaptation-related uncertainty

It is widely accepted that uncertainty pervades the whole process of adaptation to climate change, from decision-making to implementation.[23] These uncertainties must be acknowledged and managed, because they have a tendency to cause paralysis in legal and institutional frameworks.[24] Uncertainty means that governance systems cannot predict or make thorough preparation for climate change impacts, because of the imperfect knowledge of the probability, magnitude, timing, and location of such impacts.[25] We may call the main kinds of uncertainty epistemic uncertainty, ethical uncertainty, and scale uncertainty (EES). Epistemic uncertainty could be reduced by increases in knowledge through scientific research, data collection, and modelling.[26] Examples include uncertainties associated with the extremes of climate change, the vulnerability of regions, and the conditions that influence vulnerability.[27] Ethical uncertainty is about who gains and who loses from adaptation strategies, what trade-offs to make between mitigation and adaptation, and how to give a voice to vulnerable or marginalized populations.[28] Because adaptive capacity depends on a wide variety of social, political, economic, technological, and institutional factors from the country level down to the individual one, scale uncertainty exists with regard to the level at which adaptation measures should be taken.[29] EES uncertainty has been characterized in the

22 Roger Pielke et al., 'Climate Change 2007: Lifting the Taboo on Adaptation' (2007) 445 *Nature* 597.

23 McDonald (n 8); Bernie O'Brien and Mark Sculpher, 'Building Uncertainty into Cost-Effectiveness Rankings, Portfolio Risk-return Tradeoffs and Implications for Decision Rules' (2000) 38 *Medical Care* 460; John H. Mattews and A. J. Wickel, 'Embracing Uncertainty in Freshwater Climate Change Adaptation: A Natural History Approach' (2009) 1(3) *Climate and Development* 269–279; Xia Jun, Liu Chunzhen, and Ren Guoyu, 'Opportunity and Challenges of the Climate Change Impact on Water Resources' (气候变化对我国水资源影响研究面临的机遇与挑战) (2011) 26(1) *Advances in Earth Research* (地球科学进展) 1–12.

24 McDonald (n 8).

25 Anne Leitch, Ben Harman, and Marcus B. Lane, 'From Blueprint to Footprint: Climate Change and the Challenge for Planning', in Tim Bonyhady, Andrew Macintosh, and Jan McDonald (eds), *Adaptation to Climate Change: Law and Policy*, 68–69 (Federation Press, 2010).

26 J. C. Refsgaard, 'The Role of Uncertainty in Climate Change Adaptation Strategies: A Danish Water Management Example' (2013) 18 *Mitigation and Adaptation Strategies for Global Change* 337–359.

27 J. W. Handmer, S. Dovers, and T. E. Downing, 'Social Vulnerability to Climate Change and Variability' (1999) 4 (3–4) *Mitigation and Adaptation Strategies for Global Change* 267–281.

28 W. Neil Adger et al., 'Moral Reasoning in Adaptation to Climate Change' (2017) 26(3) *Environmental Politics* 371–390.

29 Katharine Vincent, 'Uncertainty in Adaptive Capacity and the Importance of Scale' (2007) 17(1) *Global Environmental Change* 12–24.

literature.[30] It produces uncertainty as to the cost, implementability, and effectiveness of any anticipatory adaptation response.[31]

Such uncertainties have a significant impact on legal and institutional frameworks, from legislative objectives to adaptation-related liability. They lead to information gaps in regulatory decision-making.[32] Information, where it exists, is not always in a useful form,[33] causing difficulty in translating scientific information into legal requirements or prohibitions. The progress of adaptation activities in reducing vulnerability is difficult to track. Quantifying loss and damage caused by adaptation failures and proving causality between actions and consequences are also difficult and at times impossible. Who is more qualified to make adaptation decisions is not always easy to decide. Supposed victims of climate change, such as those who suffer damage from adaptation failure, cannot easily establish causality to receive compensation. The list of such difficulties is almost endless.

Legal techniques help people cope with scientific uncertainty in environmental law.[34] They could be employed to deal with adaptation-related uncertainties. For example, when deciding whether adaptation measures are necessary in the face of certain risks, precautionary reasoning could be used.[35] When implementing rules on adaptation, uncertainties in implementation could be reduced through transparency, compliance management, and technical and financial assistance.[36] Jones identifies several categories of regulatory strategies, besides precaution, for dealing with scientific uncertainty in environmental regulation, including acknowledging scientific uncertainty, seeking consensus, and using the technique of adaptive management.[37]

Legal challenges presented by the context-specific nature of adaptation

Climate change impacts, vulnerability, and adaptive capacity are mostly observed and understood at a regional or local level.[38] Extreme weather events are experi-

30 Jan Rotmans, 'Global Change and Sustainable Development: A Modelling Perspective for the Next Decade' (GLOBO Report Series No. 4, Global Dynamics and Sustainable Development Programme, 1994) 20.

31 Jon Barnett, 'Adapting to Climate Change in Pacific Island Countries: The Problem of Uncertainty' (2001) 29(6) *World Development* 977, 981.

32 Robert L. Fischman and Jillian R. Rountree, 'Adaptive Management', in Michael B. Gerrard and Katrina Fischer Kuh (eds), *The Law of Adaptation to Climate Change: US and International Aspects,* 25 (American Bar Association, 2012).

33 Judith Jones, 'Regulatory Design for Scientific Uncertainty: Acknowledging the Diversity of Approaches in Environmental Regulation and Public Administration' (2007) 19(3) *Journal of Environmental Law* 347–365.

34 Jorge E. Vinuales, 'Legal Techniques for Dealing with Scientific Uncertainty in Environmental Law' (2014) 43 *Vanderbilt Journal of Transnational Law* 437–503.

35 Ibid.

36 Ibid.

37 Jones (n 33).

38 Andrew Macintosh, 'A Theoretical Framework for Adaptation Policy', in Tim Bonyhady, Andrew Macintosh, and Jan McDonald (eds), *Adaptation to Climate Change: Law and Policy,* 41 (Federation Press, 2010).

enced locally, and local knowledge is necessary to understand and manage them. Different groups of people will experience different impacts. Communities that have been marginalized because of their socio-economic status, age, gender, or ethnicity will be more vulnerable than others,[39] due to limited access to resources, technology, information, transportation, and other sustaining systems.[40] Adaptation entails a multilevel, polycentric, and non-hierarchical governance structure, as opposed to one that is state-centric and top-down.[41] Cooperation and communication among government agencies at central and local levels and affected communities are necessary for effective adaptation governance.[42] Given China's top-down political system, adaptation in China thus faces institutional challenges.

Under China's authoritarian regime, the central government has superiority over local governments in allocating targets, running a cadre evaluation system, and issuing a variety of directives.[43] It also decides many of the funding priorities for adaptation, setting up institutional structures, developing a regulatory framework, creating incentives for adaptation, and providing policy direction to lower government levels.[44] A hierarchical system that favours command-and-control regulations has been established among various levels of government (from central to provincial to prefectural to county).[45] Local governments, for their part, often need to deal with conflicts between implementing central directives and delivering on local interests.[46] The top-down regime creates obstacles to adaptation, as the latter is highly dependent on localized responses within a cooperative framework. A devolution of discretion and flexibility to local governments to develop local adaptation strategies has been called for.[47] Due to their proximity to localized climatic risks and stakeholders, local governments are in a better position to distribute climatic risks, translate adaptation goals and policies to local communities, and implement adaptation initiatives. However, reforming the

39 McDonald (n 8).

40 Federica Climato and Michael Mullan, 'Adapting to Climate Change: Analysing the Role of Government' (Evidence and Analysis Series, Department for Environment Food and Rural Affairs, 2010) 18.

41 Andrew Jordan et al., 'Emergence of Polycentric Climate Governance and Its Future Prospects' (2015)5 *Nature Climate Change* 977–982.

42 Jan Corfee-Morlot, 'Cities, Climate Change and Multilevel Governance' (OECD Environmental Working Papers, No. 14, 2009) 85.

43 He Xiangbai, 'Chinese Local Governments to be Held Responsible for Their Local Environment: New Law, Old Problems' (2019) 17(3) *China: An International Journal* 28–51.

44 Wang Hui, 'Local Level Climate Legislation in the US and Its Implications for China' (美国地方气候变化立法及其启示) (2017) 17(1) *Journal of China University of Geosciences (Social Science Edition)* (中国地质大学学报(社会科学版)) 56–64.

45 Alex Wang, 'The Search for Sustainable Legitimacy: Environmental Law and Bureaucracy in China' (2013) 37(2) *Harvard Environmental Law Review* 367–438.

46 Ibid.

47 Danny Marks, 'China's Climate Change Policy Process: Improved but still Weak and Fragmented' (2010) 19(67) *Journal of Contemporary China* 971.

current governance patterns could be challenging given China's embedded political and institutional constraints.[48]

Overview of China's policy and laws relevant to adaptation

Adaptation in Chinese policies and plans

The UNFCCC requires all state parties to formulate, implement, publish, and regularly update national programmes to mitigate climate change and facilitate adaptation.[49] As one of the measures to carry out this obligation, China's State Council promulgated its first national programme on climate change, the National Climate Change Programme (CNCCP), on 30 May 2007. The CNCCP sets out the guidelines, principles, objectives, and institutional framework to deal with climate change. It identifies the key sectors for adaptation and key measures to enhance adaptive capacity.[50] It also emphasizes the equal importance of adaptation and mitigation and the necessity to integrate adaptation into related policies and programmes.[51] In 2008, China released its first White Paper on Climate Change – titled China's Policies and Actions for Addressing Climate Change (CPAACC).[52] Since then, progress reports on the CNCCP have been published annually.[53] Since 2008, China has also initiated the development of provincial-level climate change programmes. As of November 2011, all 31 provinces (including autonomous regions and municipalities) had formulated provincial-level climate change programmes and had proceeded to implement them.[54]

Both the Twelfth and Thirteenth Five-Year Plans for national economic and social development require climate change factors to be taken into consideration in economic planning decisions and the construction of large-scale infrastructure (and other) projects.[55] In its Second National Communication,

48 Liu Shen and Zhang Yingtao, 'On the Principle of Subsidiarity and the Reform of the Chinese Administrative System' (辅助性原则与中国行政体制改革) (2006) 4 *Administrative Law Review* (行政法学研究) 9–15.

49 United Nations Framework Convention on Climate Change, opened for signature 9 May 1992, 1771 UNTS 107 (entered into force 21 March 1994) art. 4, 1(b).

50 NDRC, China's National Climate Change Program (中国应对气候变化国家方案) NDRC, Order No. 17, June 2007, 23–25.

51 Ibid.

52 Information Office of the State Council, China's Policies and Actions for Addressing Climate Change, 2008 (中国应对气候变化的政策与行动(2008)), October 2008.

53 The latest annual progress report was released in November 2018. See Information Office of the State Council, China's Policies and Actions for Addressing Climate Change, 2018 (中国应对气候变化的政策与行动(2018)), October 2018.

54 Information Office of the State Council, China's Policies and Actions for Addressing Climate Change, 2011 (中国应对气候变化的政策与行动(2011)), November 2011.

55 National People's Congress, Twelfth Five-Year Plan for National Economic and Social Development (国民经济和社会发展第十二个五年规划), 13 March 2011, Chapter 21; National People's Congress, Thirteenth Five-Year Plan for National Economic and Social Development (国民经济和社会发展第十三个五年规划), 17 March 2016, Chapter 46.

completed in 2012, China committed to developing a national adaptation policy framework.[56] Shortly thereafter, the 2013 National Strategy for Climate Change Adaptation, which is the first national adaptation policy, outlined expectations for enhancing risk management, improving relevant technical standards, and enhancing disaster-response systems.[57] It also provided a framework for provincial- and prefectural-level adaptation planning, by identifying 14 provinces for pilot projects.[58] To implement the requirements of this policy, the Urban Climate Adaptation Action Plan was co-formulated by the National Development and Reform Commission and the Ministry of Housing and Urban-Rural Development in 2016 to improve adaptive capacity by integrating adaptation factors into urban and rural planning, building-construction standards, and industrial development planning.[59] The Action Plan set a target of launching 28 pilot projects of resilient cities. The Pilot Project on the Construction of Climate-Resilient Cities, released in 2017, constitutes another policy aimed at incorporating adaptation into the whole process of urban planning and improving the climate resilience of cities.[60]

The National Plan to Address Climate Change (2014–2020), released in 2014, specifies the guiding principles on adaptation to climate change on a national level by 2020. It uses target-responsibility and accountability mechanisms to evaluate cadre performance in responding to climate change.[61] China's Third National Communication to the UNFCCC, published in 2018, clarifies the country's adaptation goals for water, agriculture, and the coastline through to 2020.[62] Pursuant to the Paris Agreement's requirement that states develop national adaptation plans, determine nationally prioritized adaptation actions, and implement relevant monitoring and evaluation mechanisms, China submitted a Nationally Determined Contribution (NDC) that details its commitment to climate change adaptation for the post-2020 period. Although some of the adaptation plans and measures in the NDC are taken from pre-existing national planning documents, their presence in the NDC gives them a higher profile and leads to greater transparency on adaptation at the international level.

56 State Council, China's Second National Communication on Climate Change (气候变化第二次国家信息通报), 2012.

57 NRDC, China's National Strategy for Climate Change Adaptation (国家适应气候变化战略), NRDC Order No 2252(2013).

58 Li Shuo et al., 'Main Focus of Adaptation under the Implementation Rules of the Paris Agreement and China's Future Measures' (《巴黎协定》实施细则适应议题焦点解析及后续中国应对措施) (2019) 15(4) *Climate Change Research* (气候变化研究进展) 436–444.

59 NRDC, City Action Plan on Climate Change Adaptation (城市适应气候变化行动方案), NRDC Order No. 245 (2016).

60 NRDC, Notice about Distributing the Pilot Work of Building Climate-resilient Cities (关于印发气候适应型城市建设试点工作的通知), NRDC Order No. (2017)343.

61 NRDC, National Plan to Address Climate Change (2014–20) (国家应对气候变化规划(2014–20)), NRDC Order No. (2014)2347.

62 MEE (n 4).

Adaptation in laws and regulations

As noted earlier, adaptation in China still lags behind mitigation in terms of its legal development, political leadership, and resource allocation.[63] In 2012, the Chinese government drafted the Law on Addressing Climate Change.[64] It required both the central and the local governments to include climate change adaptation in their national and specialized planning for economic and social development; however, it did not propose any supporting institutions or legal instruments for its implementation. (The draft law has not gone before the legislature.) Currently, no law exists whose substantive regulatory objectives include adaptation to climate change or a provision for adaptation needs.[65] Of course, "adaptive laws" containing measures that are responsive to environmental or other changes or mechanisms that enhance the flexibility and responsiveness of the law do exist.[66] But Chinese laws that seek to directly address climate adaptation challenges do not exist at present.

Two provincial regulations on climate change, in Shanxi and Qinghai, have been enacted as pilot projects. Both acknowledge the localized nature of climate change impacts and require consideration of the adverse impacts of climate change when formulating plans on land use, environmental protection, and water use.[67] Initiating legislation at the local level and then extending it to the national level is a common practice in China, and it could be justified in this case, where adaptation theories and practices are not well developed. Nonetheless, these legislative efforts are limited.

Assessment of China's adaptation framework

It may be seen from the discussion so far that China's policy and legal framework on climate adaptation is still in infancy, lacking specificity, commitment, and urgency.[68] The country's adaptation policies and plans fail to provide a clear definition of, or set a goal for, adaptation, which holds back implementation and supervision mechanisms.[69] Provincial adaptation policies and plans do not clarify

63 He Xiangbai, 'Legal and Policy Pathways of Climate Change Adaptation: Comparative Analysis of the Adaptation Practices in the United States, Australia and China' (2018) 7(2) *Transnational Environmental Law* 347–373.

64 Chinese Academy of Social Sciences, Statement on the Law on Addressing Climate Change (Proposal) (关于《气候变化应对法》（建议稿）的说明), 2012.

65 McDonald (n 8).

66 Ibid.

67 Qinghai Provincial Government, Measures on Responding to Climate Change in Qinghai Province (青海省应对气候变化办法), Qinghai Provincial Order No. 75, 23 July 2010; Shanxi Provincial Government, Measures on Responding to Climate Change in Shanxi Province (山西省应对气候变化办法), Shanxi Provincial Government Order No. 19, 23 August 2011.

68 Moore (n 20), 147–157.

69 Peng Sizhen et al., 'Current Status Problems and Recommendations on Climate Change Adaptation Policies in China' (中国适应气候变化政策现状、问题和建议) (2015) 25 (9) *China Population, Resources and Environment* (中国人口·资源与环境) 4.

who should be responsible for adaptation.[70] Nor do they respond to adaptation-related uncertainties. The precautionary principle, as an approach to adopt cost-effective measures to deal with scientific uncertainty, is absent from the Chinese environmental legal system. Adaptation strategies found in policies and plans are mostly vague or rhetorical, lacking clear substantive or procedural regulation. Criteria are lacking on how adaptation factors should be taken into account in decision-making processes or how adaptation actions are to be assessed. As a result, current land-use planning, urban planning, and infrastructure construction do not incorporate potential climatic risks or adaptation needs as much as they could.[71] Local governments have not been given incentives or the capacity to handle adaptation issues.[72] They regard adaptation as a nebulous concept.[73]

Integrating climate change adaptation within Chinese environmental laws: EIA as an entry point

The changing climate and its impacts come with corresponding challenges for the Chinese government on how to govern its environment and natural resources. Adaptation could involve a wide range of policies and laws for water, agriculture, fisheries, biodiversity, and so on.[74] Sectors identified as vulnerable to climate change impacts and are therefore also largely covered by environmental law. While there is no clear and implementable law or regulation on adaptation in China, environmental laws could be used to manage the risks and impacts of climate change. There are diverse pathways to deal with climate change adaptation through environmental laws.[75] Environmental impact assessment (EIA) is one of them.[76] As a regulatory instrument enforced by environmental authorities on a daily basis, EIA applies to many governmental plans and projects.[77] This section

70 Ibid.

71 Chen Siyu, 'Integrating Adaptation to Climate Change in Project-level Environmental Impact Assessment' (论将适应气候变化的要求纳入建设项目环境影响评价制度) (2019) 33(4) *Journal of Chongqing University of Technology (Social Science)* (重庆理工大学学报 (社会科学)) 27–37.

72 Peng et al. (n 69).

73 He Xiangbai, *Legal Methods of Mainstreaming Climate Change Adaptation in Chinese Water Management,* 94 (Springer, 2016).

74 Jonathan Verschuuren, 'Legal Aspects of Climate Change Adaptation', in Erkki Hollo, Kati Kulovesi, and Michael Mehling (eds), *Climate Change and the Law,* 258 (Springer, 2012).

75 Sofia Yazykova and Carl Bruch, 'Incorporating Climate Change Adaptation into Framework Environmental Laws' (2018) 48 *Environmental Law Reporter* 10334.

76 Ibid.

77 According to the EIA Law amended in 2018, plans that must go through the EIA process are (1) plans concerning the use of land and plans for constructing, developing, and utilizing drainage areas or sea areas made by relevant departments of the State Council and the local people's governments above the level of the cities and (2) special plans of industry, agriculture, animal husbandry, forestry, energy, water conservancy, communications, municipal construction, tourism, and natural resources.

will consider EIA as an entry point for incorporating climate change adaptation into Chinese environmental laws.

EIA has been recognized as an important legal tool to help Chinese decision makers achieve the goal of sustainable development.[78] It operates under the assumption that the government will make better decisions if it has all of the relevant information about a development proposal before it. After EIA was introduced to China in the 1970s, it was incorporated into a number of laws and regulations as a regulatory instrument.[79] Chinese EIA law requires certain activities to assess and minimize potential adverse environmental impacts before their commencement.[80] By identifying climate change impacts and the climate vulnerability of a plan or project to them, EIA has the potential to provide climate-proofing solutions.

The EIA legal regime was developed long before climate change became an important factor in government decisions.[81] Amendments to the EIA Law in 2016 and 2018 failed to consider climate change issues. The question therefore arises whether the current EIA Law of China could be used to facilitate the integration of adaptation factors.

In China's EIA Law, EIA focuses primarily on identifying and mitigating the proposed action's impact on the environment, rather than the impacts of environmental change on the proposed action.[82] Therefore, only those actions that have a potentially adverse environment impact are obligated to conduct an EIA. In the EIA process, the impacts of climate change are largely, but not entirely, ignored.[83] For example, during the environmental baseline investigation stage of an EIA, climate change impacts, such as rising temperature, changing precipitation, or the intensity and frequency of extreme weather events, are not assessed.[84] The proposed

78 Environmental Impact Assessment Law (EIA Law) (环境影响评价法), Standing Committee of the National People's Congress (adopted on 28 October 2002, amended on 2 June 2016 and 29 December 2018), art. 1.

79 E.g. Law on Atmospheric Pollution Prevention and Control (大气污染防治法), Standing Committee of the National People's Congress (adopted on 5 September 1987, amended on 29 August 1995, 29 April 2000, 29 August 2015, and 26 October 2018), art. 18; Water Pollution Prevention and Control Law (水污染防治法), Standing Committee of the National People's Congress (adopted on 11 May 1984, amended on 15 May 1996, 28 February 2008, and 27 June 2017), art. 34; State Council, The Regulation of Environmental Management of Construction Projects (建设项目环境保护管理条例), State Council Order No. 682, 1 October, 2017.

80 EIA Law (n 78), art. 2.

81 The legal framework of EIA started as early as 1990s. The EIA law was formulated in 2002, followed by the regulation of Planning EIA in 2009. By this time, climate change had not been considered by the Chinese legal system.

82 According to Article 1 and 16 of the EIA Law, the purpose and content of EIA is about assessing and preventing the negative impacts of plans and projects on the environment after they have been carried out.

83 Wu Jing and Zhang Yixin, 'The Discussion of Integrating Climate Change Factors in Environmental Impact Assessment' (关于中国将气候变化因素融入环境影响评价的探讨) (2011) 9 *Environmental Pollution and Control* (环境污染与防治) 91–94.

84 Chen (n 71).

action's vulnerability to, and capacity to deal with, climate change impacts is not part of the environmental baseline.[85] A reinterpretation of (or amendment to) existing EIA law to assess how proposed actions sensitive to climatic impacts could be affected is thus necessary.

In fact, Chinese EIA law allows for the possibility of considering climate change risks. It states that "the environmental assessment report should include the analysis, prediction and evaluation of likely adverse environmental impacts and measures for the prevention or reduction of the impacts".[86] This could be interpreted to cover information on climate change impacts. Moreover, the precautionary principle requires decision makers to adopt a more cautious approach to planning.[87] This could be another way to extend the EIA's application to climate change matters in China.

While a possible pathway exists for integrating climate adaptation in the EIA process, following it into practice would be challenging. Due to legislative loopholes, China's EIA system has long suffered from being a rubber stamp, where rent seeking and noncompliance are pervasive.[88] Chinese EIA law has also been criticized for allowing dramatic economic growth and pressure on the environment,[89] a low level of transparency, tokenistic public participation,[90] excessive intervention of local governments in EIA approval, and weak penalties.[91] Put another way, EIA has largely been treated as a formality rather than a functional tool to deliver better decisions.[92] These deficiencies make its extension to integrating adaptation uncertain. On the other hand, since the amendment of Environmental Protection Law in 2014 and the amendment of the EIA Law in 2016, there have been significant legal developments in correcting regulatory deficiencies, advancing public participation, and improving corporate compliance. These developments may facilitate an adaptation integration process and improve the resilience of future development.

Conclusion

In the past decade, there has been a remarkable increase in attention paid to adaptation – both theoretically and practically. The Paris Agreement provides new

85 The Ministry of Environmental Protection, Technical Guidelines for Plan Environmental Impact Assessment (General Principles) (规划环境影响评价技术导则(总纲)) HJ 130–2014, 2014.

86 EIA Law (n 78), art. 17.

87 Lee Godden and Jacqueline Peel, *Environmental Law: Science, Policy and Regulatory Dimensions,* 239 (Oxford University Press, 2010).

88 Ma Tianjie, 'China's Environment in 2015: A Year in Review' (China Dialogue, 23 December 2015) <www.chinadialogue.net/article/show/single/en/8497--China-s-environment-in-2-15-a-year-in-review> accessed 12 October 2019.

89 Wang Yan et al., 'Environmental Impact Assessment of Projects in the People's Republic of China: New Law, Old Problems' (2003) 23 *Environmental Impact Assessment Review* 543–579.

90 Zhao Yuhong, 'Public Participation in China's EIA Regime: Rhetoric or Reality?' (2010) 22(1) *Journal of Environmental Law* 89–123.

91 Zhao Yuhong, 'Assessing the Environmental Impact of Projects: A Critique of the EIA Legal Regime in China' (2009) 49 *Natural Resources Journal* 493.

92 Ibid.

elements on adaptation, as a goal in itself and in planning, implementation, transparency, and global stocktaking. China has adopted various policies and legal approaches to reduce vulnerability and improve resilience to climate change. This chapter finds that China's adaptation-related regulatory framework is, nevertheless, still in its infancy, lacking specificity, commitment, or implementable rules. In the absence of an adaptation law, Chinese environmental laws need to be pressed into service to manage adaptation needs. An example of a possible entry point to begin to integrate climate adaptation needs with existing law is the EIA procedure.

9 Towards a breakthrough in China's climate change litigation

Environmental public-interest litigation filed by NGOs

Richard Zhang Qing

Introduction

Climate change has become a global crisis. States, markets, and civil society all have been involved in responding to the climate emergency. In China, which is the largest greenhouse gas (GHG) emitter in the world,[1] in addition to government-led climate regulation, which has received an honourable mention,[2] environmental NGOs such as Friends of Nature (FON) have expressed their concern about climate change in their promotion of low-carbon lifestyles and environmental public-interest litigation (EPIL) cases.[3] Some scholars have suggested that certain recent FON-filed lawsuits represent the first climate change litigation (CCL) cases in China.[4]

As CCL has emerged in several jurisdictions, scholars such as Hilson, Markell, Ruhl, Peel, and Osofsky have put forward a broad definition of CCL, as consisting of "cases that take place in quasi-judicial contexts and that reference climate change amid a panoply of issues".[5] CCL core cases under this

1 Johannes Friedrich, Ge Mengpin, and Andrew Pickens, 'This Interactive Chart Explains World's Top 10 Emitters, and How They've Changed' (World Resources Institute, 11 April 2017) <www.wri.org/blog/2017/04/interactive-chart-explains-worlds-top-10-emitters-and-how-theyve-changed> accessed 3 November 2019.

2 A recent analysis from National Geographic concludes that China is on course to meet its Paris Agreement targets. See Kieran Mulvaney, 'Climate Change Report Card: These Countries Are Reaching Targets' (National Geographic, 19 September 2019) <www.nationalgeographic.com/environment/2019/09/climate-change-report-card-co2-emissions> accessed 24 September 2019.

3 FON, 'Annual Report 2018' (自然之友2018年报) (FON, 19 April 2019) <www.fon.org.cn/index.php?option=com_k2&view=item&id=13554:2018&Itemid=119> accessed 23 September 2019 (FON's 2018 Annual Report).

4 Zhang Guizhi, 'Results of the Eighth "Top Ten Public Interest Lawsuits in China" Came Out' (第八届"中国十大公益诉讼"评选结果出炉) (CBCGDF, 22 April 2019) <www.cbcgdf.org/NewsShow/4856/8369.html> accessed 20 September 2019.

5 Jacqueline Peel and Hari M. Osofsky, 'Climate Change Litigation's Regulatory Pathways: A Comparative Analysis of the United States and Australia' (2013) 35 *Law & Policy* 150, 153, 176, citing Chris Hilson, 'Climate Change Litigation in the UK: An Explanatory Approach (or Bringing Grievance Back In)', in F. Fracchia and M. Occhiena (eds), *Climate Change: La Riposta del Diritto,* 421 (Editoriale Scientifica, 2010); David Markell and J. B. Ruhl, 'An Empirical Survey of Climate Change Litigation in the United States' (2010) 40 *Environmental Law Reporter* 10644.

definition are characterized by climate-change-specific arguments or judicial analysis referencing climate change.[6] Core cases are needed to facilitate the response of climate change more directly but also to raise public awareness of climate change issues. The following table shows the definitions of these interconnected concepts more clearly. These are the definitions that will be adopted in this chapter.

Concept	*Definition*	*Example*
CCL Core Cases	Cases that include climate-change-specific arguments or judicial analysis referencing climate change (i.e. mention climate-change-related terms). They could be private or public lawsuits.	Tort-based GHG-emission cases against "carbon majors" on the control of GHG emissions (the plaintiff alleges that the defendant's GHG emissions have exacerbated climate change and caused damage to private or public interests); or administrative cases against governments for better climate regulation (the plaintiff alleges that the government has not performed its legal duty in this field and has exacerbated climate change, which has caused damage to private or public interests).
CCL-Related Cases	Cases that reference climate change amid other issues but do not address climate-change-related concerns as such. Nevertheless, these cases have positive implications for climate mitigation or adaptation (indirect and incidental effect). Both CCL Core Cases and CCL-Related Cases count as CCL cases according to Peel and Osofsky's definition.	Contractual-dispute cases involving energy enterprises (the contract at issue includes climate-change-related terms).
CCL "Indirect" Cases	Cases that do not use climate-change-related terms but that have positive implications for climate mitigation or adaptation (indirect and incidental effect).	Cases against forest destruction on the basis of a requirement of ecological conservation.

6 Peel and Osofsky (n 5), 153.

To date, according to a Guiding Document (hereinafter Opinion 2016) of China's Supreme People's Court (SPC),[7] there are four kinds of "judicial measures to respond to climate change" in this country:

1 cases on "carbon emissions"
2 cases on "energy conservation"
3 cases on "green finance"
4 cases on "biodiversity conservation".

A "Guiding Document" (or "Guideline") in this chapter refers to a policy issued by the State Council (the central government of the People's Republic of China [PRC]) or its relevant ministries (for climate change, they are mainly the Ministry of Ecology and Environment and the Ministry of Justice) or the SPC to guide or support the application of the law. Such guidelines are not legally binding, but they do influence subordinate authorities (e.g. local courts and local bureaus of ecology and environment). Opinion 2016 requires courts to hear the four kinds of cases listed, according to the law, in order to develop China's climate-governance system.[8]

A group of researchers from Sichuan University examined 177 concluded lawsuits from 2011 to 2018 falling into the four categories of Opinion 2016.[9] They found that none of these cases addressed climate-change-related concerns as such, and most of them were private, contract-based, litigation suits.[10] As per the foregoing definitions, the study's 177 cases are not sufficiently on point to be called CCL Core Cases, because they do not include climate-change-specific arguments or judicial analysis referencing climate change, even though some of them mention climate-change-related terms (e.g. a real-estate sales contract dispute referencing low-carbon policy requirements[11]).

Given the subservient relationship of the judiciary to the executive in China, it is most unlikely that CCL involving public-authority defendants will ever take off in the country unless government duties to reduce GHG emissions are prescribed by law.[12] Those who claim climate change damages, and NGOs with EPIL standing (acting as environmental public-interest representatives), may file environmental

7 Opinions of the SPC on Fully Developing Adjudication's Role and Effects to Provide Judicial Service and Safeguards for Promoting Eco-civilisation Construction and Green Development (最高人民法院关于充分发挥审判职能作用为推进生态文明建设与绿色发展提供司法服务和保障的意见), SPC Order No. (2016) 12, 26 May 2016, paras. 14–16.

8 Ibid.

9 They selected 177 cases out of 7,200 relevant cases according to Opinion 2016 to form a sample. See Zhao Yue, Lyu Shuang, and Wang Zhu, 'Prospects for Climate Change Litigation in China' (2019) 8 *Transnational Environmental Law* 349, 354.

10 Ibid., 360–362.

11 Ibid., 359.

12 Zhao, Shuang, and Zhu (n 9), 365.

tort litigation (including civil EPIL[13]) against emitters. The EPIL being filed by NGOs is a kind of emerging tort-based litigation targeting mainly private defendants (civil EPIL), because NGOs do not have a clear legal basis to bring administrative EPIL against the government.[14] Compared to procuratorates, whose court filings focus on administrative EPIL,[15] environmental NGOs are playing a central role in carrying out civil EPIL related to air pollution.[16]

Since the EPIL filed by NGOs can aim at GHG emitters directly and has already produced results in air-pollution control, the question arises whether this kind of litigation can include climate-change-specific arguments or judicial analysis to address climate-change-related concerns directly (especially GHG emissions). In other words, can it instigate the development of CCL Core Cases in China?

To answer this question, this chapter will first introduce the EPIL filed by NGOs in China and typical CCL-Related Cases and CCL "Indirect" Cases filed by NGOs. It will then analyse the challenges faced by CCL under China's existing legal framework. It also examines two recent cases filed by the NGO FON. These two suits, which object to the curtailment of renewable energy, have climate change content.[17] I ask whether they represent a kind of breakthrough. Also, in light of CCL practice, I will discuss the challenges faced by these cases and offer some directions for improvement.

Overview of the EPIL filed by NGOs in China

Although several Chinese cities had been authorizing EPIL cases filed by NGOs on a trial basis,[18] the action was not formally established until January 2015.[19]

13 Interpretation of the SPC on Several Issues of the Application of Law in the Trial of Disputes over Liability for Environmental Torts (最高人民法院关于审理环境侵权责任纠纷案件适用法律若干问题的解释), SPC Interpretation No. (2015) 12, 1 June 2015, art. 18 (The Judicial Interpretation on Environmental Tort).

14 According to Article 25 of the Administrative Litigation Law (2017 Revision), a plaintiff other than a procuratorate must show a direct interest in the administrative action at issue. As a result, only procuratorates can initiate administrative EPIL. NGOs can file only normal administrative lawsuits on the basis of their direct interests.

15 In 2018, Chinese procuratorates handled 4,393 civil public interest litigation (PIL) cases, while the number of administrative PIL cases is 108,767 (59,312 of these 110,000-odd cases are related to environmental protection). See 'Zhang Jun: Explore PIL According to the Law; Fulfil the Lofty Mission of Protecting Public Interests' (张军： 依法开拓公益诉讼，履行公益保护崇高使命) *Xinhua Net* (12 March 2019) <www.xinhuanet.com/politics/2019lh/2019-03/12/c_1210079657.htm> accessed 10 September 2019 (2018 Annual Report of the SPP). Zhang Jun is the Chief Procurator of the Supreme People's Procuratorate (SPP).

16 Zhao, Lyu, and Wang (n 9), 369.

17 E.g. FON's 2018 Annual Report (n 3).

18 For the previous legislation and practice of the trial regions, see, e.g., Bie Tao, 'The Starting Point for Legislation of EPIL – Comment and Analysis for Amendment of the Civil Procedure Law and Advice on Amendment of the Environmental Protection Law' (环境公益诉讼立法的新起点 – 《民诉法》修改之评析与《环保法》修改之建议) (2013) 1 *Law Review* (法学评论) 101, 102–104.

19 Environmental Protection Law of the PRC (中华人民共和国环境保护法), Standing Committee of the National People's Congress (NPC) (adopted on 26 December 1989, revised on 24 April 2014), art. 58. The revision came into force on 1 January 2015.

Article 58 of the revised Environmental Protection Law (the EPL 2014) empowers NGOs that meet certain conditions (including engagement in environmental protection public-interest activities for at least five consecutive years) to file litigation against environmental pollution or ecological damage to safeguard the public interest. This filled a gap left by the Civil Procedure Law (CPL).[20] Although Article 55 of the CPL prescribed that a "relevant organization" could file an EPIL case, it did not state precisely which NGOs had standing. Chinese courts have generally held that damaging the environment is tantamount to damaging the environmental public interest,[21] so environmental public interest covers the environment.

The EPIL filed by NGOs in China is paralleled by the EPIL filed by procuratorates and "ecological environmental damage compensation litigation" (EEDCL) filed by governments. Both will be discussed next.

EPIL filed by NGOs

This section will first discuss the salient features of NGO-filed EPIL. These may become an advantage or obstacle when NGO plaintiffs launch cases against climate change issues directly. I will then consider typical EPIL CCL cases filed by NGOs intended to support action on climate change.

Features of practice

By July 2019, Chinese courts had accepted 298 EPIL cases filed by NGOs under the EPL 2014 and had brought 119 of them to conclusion.[22] In prior research, I documented 214 of these 298 cases[23] and ascertained their main features.

First, in terms of initiating proceedings, NGO plaintiffs have been notably active. They tend to aim at defendants that have already been the object of administrative penalties, criminal investigations, or prosecutions.[24] But they have not been satisfied with low-hanging fruit. They have filed new types of cases, from

20 Civil Procedure Law of the PRC (中华人民共和国民事诉讼法), NPC (adopted on 9 April 1991, revised on 28 October 2007, 31 August 2012, and 27 June 2017), art. 55.

21 See, e.g., *CBCGDF* (中国生物多样性保护与绿色发展基金会) *v. Ningxia Ruitai Technology Co.* (宁夏瑞泰科技股份有限公司) [2016] Zui Gao Fa Min Zai No. 47 (Ruling on Retrial, SPC) (the Ruitai Case).

22 Dai Rui, 'SPC: 298 EPIL Cases Filed by NGOs Has Been Accepted Since 2015' (最高法：2015年以来受理社会组织提起环境公益诉讼案件298件) *The Cover* (30 July 2019) <www.thecover.cn/news/2410618> accessed 12 September 2019.

23 For a complete list of these 214 cases, see Zhang Qing, *Ground-breaking but not so Effective: Environmental Public Interest Litigation Filed by NGOs in China* (PhD Thesis, The Chinese University of Hong Kong, 2019). All of these cases were accepted before 31 December 2018. The cases accepted after 2019 are not a focus here, because they had only recently started and the relevant statistics were not fixed. The features that follow are also findings of this thesis.

24 According to the statistics of the documented cases, the defendants of at least 138 cases (more than 60%) had been investigated.

protecting immovable cultural relics to avoiding damage to the ecology by seeking to reduce risk.[25] To achieve their protection goals, and subject to the SPC's "judicial interpretation" on EPIL,[26] NGO plaintiffs have made claims borrowed from private-interest-based tort litigation, including "ceasing the infringement" and "remediating the damaged environment" (these two types account for around 70% of the cases) – while "compensating for losses of ecosystem services" and "apology" have also been common.[27]

Second, in terms of hearing lawsuits, progress relies heavily on court activism. In practice, governmental environmental agencies seldom respond to the courts' requests for documents or other information, which at times makes it difficult for the courts to investigate and decide facts. Additionally, even though expert opinion (from a person or an institution with special knowledge of a particular field,[28] including an opinion from a local governmental environmental agency)[29] plays an important role in explaining scientific issues to the court (e.g. how to appraise the environmental damage), courts regard an "appraisal opinion" (the opinion of a qualified appraisal expert or institution on specific issues,[30] such as an appraisal report on environmental damage) as a first choice for reference when deciding issues of fact, because it is more formal (the appraisal expert or institution has a qualification certificate) and objective.[31] As a result, some new-type cases are pending (and some of them have been pending for more than two years) due to the lack of a competent appraisal institution,[32] Appraisal Guideline,[33] or Hearing Guideline. If a court is proactive, it will make full use of existing mechanisms and

25 See, e.g., *China Biodiversity Conservation and Green Development Foundation (CBCGDF) v. Yalung River Basin Hydroelectric Development Co*., which aims to protect an endangered species from dam construction (the Yalung Case).

26 Interpretation of the SPC on Several Issues Concerning the Application of Law in the Hearing of Civil EPIL (最高人民法院关于审理环境民事公益诉讼案件适用法律若干问题的解释), SPC Interpretation No. (2015) 1, 7 January 2015, art. 18 (Judicial Interpretation on Civil EPIL). The SPC's Judicial Interpretation is binding, according to art. 104 of the Legislation Law (2015 Revision).

27 See fn 23.

28 Civil Procedure Law (2017 Revision, CPL 2017 hereinafter), art. 79.

29 For example, in *All-China Environment Federation (ACEF) v. Longhai Huayu Hardware Manufacturing Co. et al*., after the plaintiff applied for appraising a remediation fee, the court asked Fujian Provincial Department of Environmental Protection for an expert opinion and dealt with this issue on the basis of its reply.

30 CPL 2017, art. 76.

31 In addition, pursuant to the CPL 2017, "appraisal opinion" is an independent evidence category, while "expert opinion" is treated as a kind of "statement of a party" (which is also an independent evidence category but is relatively subjective).

32 See, e.g., *FON v. Yunnan Jiangchuan Xianhu Jinxiu Tourism and Property Development Co*.

33 A kind of Guiding Document to streamline the appraisal process – e.g. Notice of the General Office of the Ministry of Environmental Protection (MEP) on Issuing the 'Recommended Methods for Environmental Damage Appraisal During the Emergency Disposal Stage of Environmental Emergency' (环境保护部办公厅关于印发《突发环境事件应急处置阶段环境损害评估推荐方法》的通知), MEP General Office No. (2014) 118, 31 December 2014.

decide a case in a reasonable amount of time;[34] but if the court is overcautious, it will set aside the case and keep postponing the hearing date indefinitely, because there is no maximum period for hearing a civil case.[35]

Third, in terms of making and enforcing judgments, court practice is becoming more predictable, but tricky issues remain. As mentioned, courts generally hold that damaging the environment also damages the environmental public interest. Most (more than 90%) of the reviewed cases were decided in favour of the plaintiffs. Courts will refer to multiple documents when writing their judgments – from the aforementioned expert and appraisal opinions to other relevant judgments[36] and the SPC's Guiding Documents.[37] Some courts have produced detailed remediation instructions for the defendants to follow, which make monitoring possible. The affected environment in some cases has seen substantive improvement after judgment enforcement.[38] Nevertheless, managing the compensation payment by the defendants needs to be optimized to fully realize public monitoring and participation. According to the documented judgments, most compensation goes into governmental "EPIL Funds", local state treasuries, or the court's designated accounts, but the public, and even the NGO plaintiffs themselves, seldom have channels to make sure that the money is used properly for environmental remediation.

Connection with CCL

EPIL filed by NGOs enables the public to protect environmental public interests more directly through enforceable judgments and by improving law enforcement (e.g. having the alleged infringement stopped during the lawsuit, when the relevant authority has refused or failed to do so).[39] Since Chinese courts generally

34 E.g. *ACEF v. Changjing Liangping Live Pig Cooperative of Jiangyin et al.* See 'The White Paper of the Environmental and Resources Adjudication in Jiangsu' (江苏环境资源审判白皮书) (The Higher People's Court of Jiangsu Province, 5 June 2017) <www.jsfy.gov.cn/art/2017/06/05/25_91797.html> accessed 31 October 2019.

35 CPL 2017, arts. 149 and 176.

36 See, e.g., *ACEF* (中华环保联合会) *v. Xin'an Chemical* (浙江新安化工集团股份有限公司) *et al.* [2017] Lu Min Zhong No. 1577 (Higher People's Court of Shandong Province).

37 See, e.g., *ACEF* (中华环保联合会) *v. Yichun Zhongan Industry Co.* (宜春市中安实业有限公司) *et al.* [2018] Gan Min Zhong No. 189 (Higher People's Court of Jiangxi Province).

38 See, e.g., *FON* (北京市朝阳区自然之友环境研究所) *and Guangdong Environmental Protection Foundation* (广东省环境保护基金会) *v. Shenzhen Dongyangguang Industrial Development Co.* (深圳市东阳光实业发展有限公司) *et al.* [2016] Yue 18 Min Chu No. 3 (Mediation Document, Qingyuan Intermediate People's Court, Guangdong Province). On its remediation effect, see FON, 'Guangdong Nanling: the Ecological Remediation Is Ongoing: Thanks for the Steady Efforts' (广东南岭：生态修复进行时，感谢所有一步一个脚印的努力) (FON, 25 June 2018) <www.fon.org.cn/index.php?option=com_k2&view=item&id=13109:2018-06-25-03-43-27&Itemid=176> accessed 17 September 2019.

39 E.g. *FON v. Xinping Development Co. of HydroChina et al.*, a case aiming to protect the affected green peacocks, contributed to a temporary halt to the construction of a dam. See Ma Rongzhen, 'A "Green Peacock Defence Battle" Initiated by Many ENGOs' (一场多家环保NGO掀起的"绿孔雀保卫战") *Southern Weekly* (Guangdong, 30 September 2018) <www.infzm.com/content/139882> accessed 17 September 2019.

accept that climate change will have adverse effects on the environment, litigation against actions tending to exacerbate climate change may be accepted by them on the condition that the plaintiff can provide relevant evidence – such as evidence of causation and damage. Via this channel, courts could, in theory, optimize climate change regulation through supplementing, monitoring, and guiding administrative law enforcement,[40] acting as "collaborators in the regulatory process",[41] and helping implement relevant laws and policies.

Typical CCL-Related Cases and CCL "Indirect" Cases filed by NGOs

By 31 December 2018, eligible NGOs had filed several CCL-Related Cases and CCL "Indirect" Cases. In the sample of 214 cases, I counted 24 air-pollution suits concerning pollutants that also play a role in climate change (NO_X and CO)[42] and 41 ecological-damage suits involving carbon sinks (forests, lakes, natural reserves, wetlands, farmland, and endangered plant species).

Two ground-breaking decided cases illustrate the two categories:

1 *Green Home Environment-Friendly Center of Fujian and FON v. Xie Zhijin et al*. In this "first-ever civil EPIL case under the new EPL" (accepted in January 2015), the plaintiffs demanded of the defendants, who damaged forests in the course of mining activities, to remediate forest vegetation and compensate the public for the loss of ecosystem services.[43] The first- and second-instance courts both decided in favour of the plaintiffs (and the compensation would be managed by the first-instance court to improve the environment for the public).[44] In deciding the amount of environmental damage, both courts adopted the expert opinion presented by the plaintiffs.[45] The opinion included damage from "carbon release", caused by vegetation destruction, in the loss of ecosystem services.[46] *Carbon release* is clearly a term related to

40 Other than supplementing administrative law enforcement, EPIL can expose serious violations and urge the corresponding authority to perform its duties. Furthermore, the court could issue a "Judicial Recommendation" to suggest to the authority some directions for improvement. See the CPL 2017, art. 114.

41 Zhao, Lyu, and Wang (n 9), 351.

42 Although NO_X and CO are not GHGs, they can facilitate climate change through catalysing or contributing to the formation of GHGs. That is, they are reactive gases. See J. T. Houghton et al. (eds), *Climate Change 2001: The Scientific Basis,* 256, 259 (Cambridge University Press, 2001).

43 SPC, 'A Civil PIL Case on Forest Damage: FON and Green Home Environment-Friendly Center of Fujian v. Xie Zhijin and Three Other Persons' (北京市朝阳区自然之友环境研究所、福建省绿家园环境友好中心诉谢知锦等四人破坏林地民事公益诉讼案) (China Court Net, 29 December 2015) <www.chinacourt.org/article/detail/2015/12/id/1777817.shtml> accessed 18 September 2019.

44 Ibid.

45 *Xie Zhijin* (谢知锦) *et al v. FON* (北京市朝阳区自然之友环境研究所) *and Green Home Environment-Friendly Center of Fujian* (福建省绿家园环境友好中心) [2015] Min Min Zhong Zi No. 2060 (Higher People's Court of Fujian Province).

46 Ibid.

climate change. However, there was no climate-change-specific argument or judicial analysis in this case. Hence, it is a CCL-Related Case.

2 *All-China Environment Federation (ACEF) v. Dezhou Jinghua Group Zhenhua Co.* (the Jinghua Case). This was the first EPIL case concerning air pollution (accepted in March 2015).[47] ACEF claimed that the Zhenhua Company's excessive emissions had caused serious air pollution (one of the major pollutants was NO_X) and demanded infringement cessation, compensation, and an apology.[48] The court found for the NGO on the ground of administrative law-enforcement records and an appraisal opinion on damages.[49] When hearing the case, the court cooperated with the local environmental authority to ensure the cessation of the pollution and relocation of the company to a new site far from residential areas.[50] The case illustrates the judiciary's potential to optimize administrative enforcement. Because this lawsuit targets NO_X but does not mention climate change, it is a CCL "Indirect" Case.

From these two cases, we gain a sense of the indirect or incidental effects of CCL-Related Cases and CCL "Indirect" Cases. Moreover, since the EPL 2014 itself does not aim at climate change directly, and the Law on Prevention and Control of Atmospheric Pollution does not clarify specific measures to "control air pollutants and GHGs in a coordinated manner" or the party responsible for this duty,[51] even if a court were proactive[52] and wanted to carry out low-carbon policies, it may not be able to do so.

EPIL filed by procuratorates and EEDCL

In accordance with Article 55 of the CPL (2017 Revision) and Article 25 of the Administrative Litigation Law (2017 Revision), China's procuratorates, which are "legal monitoring authorities" (their mission includes safeguarding national and public interests and ensuring the correct implementation of the law through the

47 *ACEF* (中华环保联合会) *v. Dezhou Jinghua Group Zhenhua Co.* (德州晶华集团振华有限公司) [2015] De Zhong Huan Gong Min Chu Zi No.1 (Dezhou Intermediate People's Court, Shandong Province).

48 Ibid.

49 Ibid.

50 'The SPC Issued EPIL Model Cases' (最高人民法院发布环境公益诉讼典型案例) *Xinhua Net* (Beijing, 7 March 2017) <www.xinhuanet.com//legal/2017-03/07/c_129503217.htm> accessed 19 September 2019.

51 Law on Prevention and Control of Atmospheric Pollution of the PRC (中华人民共和国大气污染防治法), Standing Committee of the NPC (adopted on 5 September 1987, last amended on 26 October 2018), art. 2. This law, along with the EPL 2014 as well as its "judicial interpretations", is the most important legal basis of the EPIL cases in these two categories (CCL-Related Cases and CCL "Indirect" Cases).

52 For the meaning of *proactive courts* in this chapter, see "Features of practice", above (the second feature).

exercise of procuratorial power),[53] can file a civil EPIL case against an infringer and can sue governmental environmental agencies that are not performing their duties as prescribed by law.

As mentioned at the start of this chapter, procuratorates are more focused on administrative EPIL, and they will bring a civil EPIL case only when there is no eligible administrative organ or NGO to do so or where an eligible party does not do so.[54] Still, procuratorates have initiated more civil EPIL suits than have NGOs,[55] as a result of their higher capacity (funding, staffing, etc.) and institutional support (e.g. exemption from paying a litigation fee).[56]

Is the EPIL filed by procuratorates likely to play an important role in the development of China's CCL Core Cases? The answer may not be positive. As at the end of 2019, no CCL case filed by a procuratorate had been documented. We can identify two major obstacles. First, as some scholars have pointed out, while procuratorates could aim their suits at local governmental environmental agencies, they could hardly litigate public-authority defendants if no law prescribes a governmental duty to reduce GHG emissions.[57] And this, in fact, is the reality. Since procuratorates are legal monitoring authorities, the agencies monitored must have duties prescribed by law. Procuratorates are to urge the relevant administrative organs to correct unlawful administrative acts or perform their duties according to the law before taking them to court.[58] Second, in the field of civil EPIL, a procuratorate generally cannot file a lawsuit outside its jurisdiction, so other procuratorates cannot sue polluters in that jurisdiction if the eligible procuratorate was hindered, say, by local protectionism.[59] By comparison, NGOs are not limited by

53 Constitution of the PRC (中华人民共和国宪法), NPC (adopted on 4 December 1982, last amended on 11 March 2018), art. 134. Organic Law of the People's Procuratorates of the PRC (中华人民共和国人民检察院组织法), Standing Committee of the NPC (adopted on 1 July 1979, last revised on 26 October 2018), arts. 2 and 20.

54 CPL 2017, art. 55.

55 For example, in 2018, Chinese courts accepted 1,737 EPIL suits filed by procuratorates and 65 civil EPIL suits filed by NGOs. Of the 1,737 cases brought by prosecutors, 113 are civil EPIL (excluding the incidental civil PIL in environmental criminal action). See 'Chief Judge of the Environmental and Resources Division of the SPC Wang Xuguang Talks about the Work on Environmental and Resources Adjudication' (最高法环境资源审判庭庭长王旭光谈环境资源审判工作) *Legaldaily.com* (Beijing, 10 March 2019) <www.legaldaily.com.cn/direct_seeding/node_101925.htm?from=timeline&isappinstalled=0> accessed 21 September 2019 (Interview with the Chief Judge of SPC Environmental Division).

56 Measures for the Implementation of the Pilot Program of Hearing by People's Courts of PIL Cases Instituted by People's Procuratorates (人民法院审理人民检察院提起公益诉讼案件试点工作实施办法), SPC Order No. (2016) 6, 25 February 2016, art. 22 (Measures of Hearing PIL Filed by Procuratorates).

57 Zhao, Lyu, and Wang (n 9), 365.

58 E.g. Measures of Hearing PIL Filed by Procuratorates, art. 12.

59 Yang Kaiqi, 'EPIL Filed by Procuratorates: Nearly 60,000 Cases a Year; They Confront the Administration with Toughness but Do not Break up the Relationship' (检察院环境公益诉讼：一年近六万件，硬碰硬但不撕破脸) *Southern Weekly* (Guangdong, 22 March 2019) <www.infzm.com/content/146284> accessed 21 September 2019.

their place of registration (e.g. a Beijing NGO could file a case against a polluter in Gansu), so they can better resist local protectionism, which contributes to their central role in carrying out air-pollution civil EPIL.

An analogous procedure is offered by EEDCL cases. Pursuant to a Communist Party policy,[60] local governments or the ministry delegated by the State Council to exercise "the ownership of natural resources owned by the whole people" could file an EEDCL case against a party that has caused damage to the environment if the agency fails to reach an agreement with the alleged offender through consultation. EEDCL claims can be similar to those of civil EPIL (e.g. remediating and compensating), and EEDCL cases follow the CPL.[61] Although an EEDCL case looks very much like a civil EPIL case filed by the government, it is essentially a kind of lawsuit associated with a special procedure to protect national environmental interests, in which a government acts as a representative of state-owned natural resources.[62]

The term *ecological environmental damage* in EEDCL refers to adverse changes to environmental elements (e.g. the atmosphere) or biological elements (flora and fauna) as well as the functional degradation of ecosystems caused by polluting the environment or destroying the ecology.[63] Certain emissions or ecological damage accelerate climate change and thus bring about adverse environmental impacts. EEDCL lawsuits could include climate-change-specific arguments, given the central government's support for climate action.[64] However, this system is still relatively new (effective nationwide on 1 January 2018). By July 2019, courts had accepted only 30 such suits and finalized 25 of them.[65] No CCL case has yet appeared among them.

In summary, the procuracy could file administrative EPIL cases against governmental environmental agencies for not managing GHG emissions properly as long as their duties in this field are stipulated by law. It could also file civil EPIL cases against GHG emitters to have them cease emissions or to provide compensation where the fact and level of the damage, as well as causation, are possible to prove. Of the two, a civil EPIL case to protect a carbon sink (e.g. a forest) on the

60 Plan on the Reform of the System of Ecological Environmental Damage Compensation (生态环境损害赔偿制度改革方案), The General Office of the Communist Party of China Central Committee and the General Office of the State Council (No Document Number), December 2017, point 3 and 4.

61 Several Provisions of the SPC on the Hearing of EEDCL Cases (For Trial Implementation) (最高人民法院关于审理生态环境损害赔偿案件的若干规定（试行）), SPC Interpretation No. (2019) 8, 4 June 2019, art. 11 (Judicial Interpretation on EEDCL).

62 According to art. 9 of the PRC Constitution, "owned by the whole people" is tantamount to "owned by the state". The SPC also thinks EEDCL is different from EPIL. E.g. Interview with the Chief Judge of SPC Environmental Division (n 54).

63 Point 3 of the Plan on the Reform of the System of Ecological Environmental Damage Compensation.

64 E.g. Official Reply of the State Council on the National Plan to Respond to Climate Change (2014–2020) [国务院关于国家应对气候变化规划(2014–2020年)的批复], State Council Letter No. (2014) 126, 17 September 2014.

65 Dai (n 22).

basis of ecological conservation and climate-change-specific arguments would face fewer challenges.

Challenges for CCL in China

Since China's EPIL is growing and has made good progress, could it propel the response to climate change more directly? Before answering this question, this section will identify the most pressing challenges faced by CCL in China under the existing legal framework. If NGOs and other eligible parties hope to bring more CCL cases (especially CCL Core Cases) and succeed, these challenges will need to be dealt with.

As the earlier discussion has shown, the most formidable challenge for administrative CCL is that government duties to reduce GHG emissions are not prescribed, or clearly prescribed, by law (e.g. in the Law on Prevention and Control of Atmospheric Pollution), so the eligible party for administrative EPIL (i.e. procuratorates) does not have a substantive legal basis to sue a government agency for not managing the control of GHG emissions properly. Likewise, a victim suffering from adverse impacts allegedly caused by climate change also cannot file a normal (private-interest) administrative lawsuit against the agency.[66]

For civil CCL, there are two urgent challenges. The following paragraphs will concentrate on the challenges faced by tort-based civil CCL because, compared to contract-based CCL, it can aim at the damage caused by climate change more directly without requiring a contractual relationship between the parties.

Plaintiff's challenge: difficulty in proving emissions, damage, and causation

The first challenge lies in the proof of GHG emissions, damage, and causation. According to the CPL and its "judicial interpretations", filing civil litigation in China requires meeting four requirements:

1 The plaintiff has a direct interest in this case (e.g. the plaintiff is a victim) or has EPIL standing (e.g. the plaintiff is an eligible NGO in accordance to Article 58 of EPL 2014) or has EEDCL standing.
2 The defendant is specific (the plaintiff can specify who the defendant is).
3 The claims, as well as their supporting facts and reasons, are specific (not abstract).
4 The dispute is in the range of civil action, and the court has jurisdiction.[67]

66 Administrative Litigation Law of the PRC (中华人民共和国行政诉讼法), NPC (adopted on 4 April 1989, revised on 1 November 2014 and 27 June 2017), art. 12.

67 CPL 2017, arts. 55 and 119; Judicial Interpretation on Civil EPIL, art. 1; Judicial Interpretation on EEDCL, art. 5.

Moreover, in environmental tort litigation, the plaintiff shoulders the initial burden of proof, consisting of the following elements:

1. The polluter emitted the pollutant.
2. There was damage (in the EPIL, it is damage or a significant risk of damage to the public interest).
3. There was causation between the pollutant and the damage.[68]

If the plaintiff manages to prove these three elements, it is the defendant's turn to provide evidence against the plaintiff's charge.[69] There are three possible scenarios at this point: first, the defendant fails to do so, and there is no circumstance of mitigation or exemption from liability, in which case the defendant shoulders tort liability; second, the defendant provides persuasive counterevidence (e.g. on the absence of causation), in which case the plaintiff may be allowed to reinforce its evidence (e.g. by presenting another expert opinion) and start a new round of hearings; and third, the defendant's evidence convinces the court or the plaintiff fails to reinforce its previous evidence, in which case the plaintiff loses.

According to the SPC, the environmental tort plaintiff has a relatively light burden of proof regarding causation – it has to establish only that the pollutant is related to the damage;[70] however, the Tort Law and the SPC's "judicial interpretation" on environmental tort do not specify how to establish that "the pollutant is related to the damage". In CCL cases against carbon emitters, the plaintiff may face difficulties when collecting evidence to prove the defendant's GHG emissions, because it may not be able to directly acquire emission information from the relevant enterprise. Unlike the relatively complete disclosure of air-pollutant emissions,[71] an official disclosure system for GHG emissions is still under development. Only a handful of provinces (e.g. Sichuan[72]) have worked out detailed

68 Tort Law of the PRC (中华人民共和国侵权责任法), Standing Committee of the NPC (adopted on 26 December 2009), art. 65; Judicial Interpretation on Environmental Tort, arts. 1 and 6; Judicial Interpretation on Civil EPIL, art. 1.

69 E.g. Judicial Interpretation on Environmental Tort, art. 7.

70 Sun Ying, 'The SPC Issued the Judicial Interpretation on Environmental Tort' (最高人民法院发布环境侵权司法解释) *China National Radio* (Beijing, 1 June 2015) <http://china.cnr.cn/gdgg/20150601/t20150601_518706638_1.shtml> accessed 3 December 2019. This report quotes the SPC's opinion on the burden of proof of environmental tort.

71 Under instructions like the 'Measures for the Self-monitoring and Information Disclosure by the Enterprises Subject to the Intensive Monitoring and Control of the State (for Trial Implementation)' (国家重点监控企业自行监测及信息公开办法(试行)), specific enterprises must disclose their self-monitoring of air-pollutant emission information through the platform established by the corresponding local governmental environmental agencies. Actually, some NGOs, such as the Institute of Public & Environmental Affairs (IPE), have compiled such data (wwwen.ipe.org.cn/MapPollution/Pollution.aspx?q=3&type=1), and the public can search them conveniently.

72 Notice of the Sichuan Provincial Development and Reform Commission (SDRC) on Carrying out the GHG Emission Information Disclosure Work of Enterprises and Public Institutions in an Orderly Manner (四川省发展和改革委员会关于有序开展企（事）业单位温室气体排放信息披露工作的通知), SDRC Environment and Resources No. (2018) 91, 24 February 2018.

arrangements to carry it out. In most cases, the plaintiff could only roughly work out the emissions (e.g. with the help of an expert), which is less convincing than using governmentally approved data.

Even when the GHG-emission information is available, it may be challenging for the plaintiff to prove damage and causation. In private-interest cases, although damage to personal and property rights is possible to appraise (e.g. through a medical or asset-evaluation report), there is no environmental damage appraisal institution that can analyse the causation between a defendant's GHG emissions and the alleged harm.[73] For public-interest cases (e.g. the Jinghua Case) on air pollution, the plaintiff could prove that the defendant has exceeded the applicable pollutant-emission standard, and thus convince the court that the polluter has damaged the environmental public interest (by damaging the environment). Appraisal Guidelines issued by the Ministry of Environmental Protection (MEP; renamed Ministry of Ecology and Environment in 2018[74]) could also be used to evaluate damage by air pollution,[75] so the court could impose civil liability (tort liability) on the defendant. However, there is no law stipulating the liability caused by "excessive" GHG emissions and no GHG-emission standard[76] or Guideline on appraising the environmental damage caused by a given quantity of GHG emissions. As a result, because the defendant does not have administrative or criminal liability for its alleged excessive GHG emissions, the plaintiff is unlikely to get an appraisal opinion on the defendant's damage or any related law-enforcement record. Since the causation between the emissions and the damage is always complex, the court would not find for the plaintiff without a benchmark or evidence from law-enforcement records or appraisal reports. Therefore, compared to private-interest air-pollution lawsuits or air-pollution EPIL, the plaintiff of tort-based CCL on GHG emissions faces great difficulties and will probably lose.

In the case of tort-based CCL on carbon-sink protection, the hurdles would be less frustrating. Take civil EPIL on forest destruction as an example. The evidence on the defendant's tortious conduct, the environmental damage, and the causation between the two are not difficult to collect, because law-enforcement records from forestry authorities (showing that the defendant has destroyed a certain forest)

73 There are institutions that can study the causation between the defendant's air-pollutant emissions and the harm. The 'Rules for the Registration and Review of Environmental Damage Judicial Appraisal Institutions' (环境损害司法鉴定机构登记评审细则), which became effective on 14 June 2018, do not include the causation analysis of climate change damage in their "professional competence requirements". Nevertheless, they include the causation analysis of air pollution, so it is possible for air-pollution victims to find a competent appraisal institution.

74 Decision of the First Session of the Thirteenth NPC on the State Council Institutional Reform Plan (第十三届全国人民代表大会第一次会议关于国务院机构改革方案的决定), NPC (adopted on 17 March 2018), Point 1. Local environmental protection bureaus have been renamed bureaus of ecology and environment since 2018.

75 E.g. Notice of the General Office of the MEP on Issuing the 'Recommended Methods for Environmental Damage Appraisal During the Emergency Disposal Stage of Environmental Emergency'.

76 In the standard database of the Ministry of Ecology and Environment (http://bz.mee.gov.cn), currently there is no national standard containing "GHG" (温室气体) in its name.

and appraisal reports (showing the value of the destroyed forest)[77] are reasonably available. After all, forest destruction will constitute administrative or criminal liability in China.[78] Therefore, the court is able to determine the defendant's tort liability according to the law-enforcement record (establishing tortious conduct and causation) and the appraisal report (evaluating environmental damage). It is possible to appraise the amount of carbon that a forest stores,[79] which makes quantifying the environmental damage of losing a sink more accurate. This of course is not damage *from* GHGs but damage leading to the *release* of GHGs.

To summarize, the challenge in proving GHG emissions and their effects results, in the first place, from the difficulty in collecting persuasive evidence of the defendant's emissions. Secondly, while the damage to private or public interests could be evaluated, causation between the emissions and the damage is not currently provable, because there is no appraisal institution or relevant benchmark in this sphere. It would in any case be nearly impossible to link given emissions with localized damage, as climate change acts on a global scale.

Court's challenge: no guidelines on hearing CCL

In light of the gaps in GHG-emission regulation, it is necessary to develop relevant guidance for courts or agencies to follow. Initiating the development of guidelines on emerging litigation is feasible, given that the SPC has already done something similar before. In 2017, it issued a document on hearing EPIL,[80] which includes detailed guidance for its stages, from acceptance[81] to enforcement. However, the challenge is the lack of a Court Guideline on hearing CCL, especially the tricky CCL cases on GHG emissions. Lack of a guideline may make such cases time-consuming or unlock a court's unpredictable activism – which is like what happened to other new-type cases.[82]

77 For example, as per its judgment, the evaluation fee of *Green Home Environment-Friendly Center of Fujian and FON v. Xie Zhijin* (the Xie Zhijin Case) is RMB6,000 (the evaluation report was regarded as an expert opinion by the courts). Most NGOs with EPIL standing have RMB100,000 to RMB1,000,000 to spend annually. See Zhang (n 23).

78 Forest Law of the PRC (中华人民共和国森林法), Standing Committee of the NPC (adopted on 20 September 1984, last amended on 27 August 2009), art. 39; Criminal Law of the PRC (中华人民共和国刑法), NPC (adopted on 1 July 1979, last amended on 4 November 2017), arts. 342 and 345.

79 E.g. 'Carbon Sink Evaluation' (碳汇评估) (Beijing Zhonglin Assets Appraisal Co., 19 December 2014) <www.zhl-cpv.com/html/2014120975.html> accessed 4 November 2019. In the Xie Zhijin Case, Beijing Zhonglin Assets Appraisal Co. was delegated by the plaintiffs and presented the expert opinion on the environmental damage.

80 Work Rules of the SPC on Hearing EPIL (For Trial Implementation) [最高人民法院关于审理环境公益诉讼案件的工作规范（试行）], SPC (no Document Number), 1 April 2017.

81 Accepting a case in China means that the court agrees to register the case and launch the hearing process formally. In Chinese, *case acceptance* is equivalent to 案件受理 and *registering a case* is equivalent to 立案. See CPL arts. 123 and 125.

82 See "Features of practice", above.

The first guideline that needs to be developed in this field should be related to tort-based CCL on GHG emissions, because, first, administrative CCL on GHG emissions lacks legal bases to decide the defendant's nonfeasance or misfeasance, and a Guiding Document cannot replace the law to create a legal basis; second, contract-based CCL focuses on contractual disputes and is a less direct channel to deal with climate change concerns; third, CCL on carbon-sink protection, whether civil or administrative, faces fewer challenges due to the clear legal bases on forest protection; and fourth, normally, tort-based CCL on GHG emissions will target energy or industry emitters, because these play a dominant role in China's GHG emissions;[83] also, it has a basic statutory framework which is provided for by existing rules, such as the tort law and its "judicial interpretations". The prospective guideline could clarify the applicable range, verification of emission-related conduct (including emitting GHGs directly and facilitating the emission of GHGs), damage evaluation, causation identification (applying the "light" burden of proof, as interpreted by the SPC), and remedies.

Recent CCL practice in China and directions for improvement

So far, only two CCL Core Cases have been filed by NGOs: *FON v. State Grid Gansu Electric Power Co.* (the Gansu Grid Case) and *FON v. State Grid Ningxia Electric Power Co.* (the Ningxia Grid Case). Both were still pending at the time of writing.

The Gansu and Ningxia Cases

In the Gansu Grid Case (accepted between June and August 2017) and the Ningxia Grid Case (accepted in January 2018), the NGO plaintiff contends that the grid companies failed to purchase the qualified electricity generated by wind power and photovoltaic power in Gansu and Ningxia in the full amount in accordance with the Renewable Energy Law (REL), causing this wasted renewable electricity to be replaced by coal power.[84] The plaintiff requested that the defendants:

1 Pay for the wasted renewable electricity.
2 Compensate for the environmental damage caused by burning the substitute coal (FON estimates that compensation would cost at least RMB310,000,000 in the Ningxia Grid Case).
3 Issue a public apology.[85]

83 The emissions from the energy and industry sectors account for more than 92% of China's GHG emissions. See Friedrich, Ge, and Pickens (n 1).

84 Diao Fanchao, 'NGO Claims RMB 0.3 Billion for Wind and Solar Curtailment; Ningxia Grid: We Are Not Able to Purchase the Electricity in Full Amount' ("弃风弃光"被环保组织索赔3亿，宁夏电网：做不到全额收购) (The Paper, 11 April 2018) <www.thepaper.cn/newsDetail_forward_2069105> accessed 10 November 2019; FON's 2018 Annual Report (n 3).

85 Ibid.

FON submits that coal power produces not only air pollutants but also a great amount of GHGs, which cause global climate change and imperil the environment, so the defendants' nonfeasance has harmed public interests and should lead to civil liability.[86] Additionally, FON asserts that the Chinese government has strongly backed the development of clean energy in order to combat climate change,[87] but a large amount of wind and solar curtailment exists (the curtailment rate of wind power in Gansu was as high as 43% in 2016).[88]

Given their clear climate change framing and climate-change-specific arguments, these two related cases are CCL Core Cases. Although they mention "air pollutants", the substantive legal basis of these two cases is the REL. Wang Canfa has called the Gansu Grid Case "the first climate change law case in China".[89]

The Gansu Grid Case was dismissed by the first-instance court (Lanzhou Intermediate People's Court had jurisdiction at that time) in August 2018 for the stated reason that the grid company itself could not cause pollution to the environment.[90] No court trial was held. FON appealed, and in December 2018, the appellate court reversed the original ruling, citing the aforementioned conditions for filing a civil EPIL case (including that the plaintiff has EPIL standing)[91] and ordered Gansu Kuangqu People's Court (which has now assumed jurisdiction for Gansu's civil EPIL) to hear it.[92]

While the first- and second-instance courts both made procedural decisions ("rulings"),[93] the first-instance court ruled against the plaintiff on substantive grounds, whereas the second-instance court supported the plaintiff on the basis of procedural grounds. Another observation is that the ruling of the first-instance court is questionable because the substantive issue (including whether

86 Ibid.

87 FON, 'Gansu Wind Curtailment Case: We Start Again' (甘肃弃风案，我们"重头再来") (FON, 28 January 2019) <www.fon.org.cn/index.php?option=com_k2&view=item&id=13433:2019-01-28-03-10-35&Itemid=176> accessed 7 December 2019 (FON's Update on the Gansu Grid Case).

88 These data, which are adopted by FON, are from China's National Energy Administration (NEA). According to the NEA, the amount of nationwide wind power curtailment was 49.7 billion kWh in 2016. Nevertheless, the NEA and FON did not compare them with the normal level of curtailment. See 'Operational Condition of Wind Power Connecting to Grid in 2016' (2016年风电并网运行情况) (NEA, 26 January 2017) <www.nea.gov.cn/2017-01/26/c_136014615.htm> accessed 8 November 2019.

89 Zhang (n 4).

90 Diao Fanchao, 'Lanzhou Intermediate People's Court Dismissed the Wind and Solar Curtailment Case FON v. State Grid Gansu Electric Power Co.: The Litigation Is not Eligible' (兰州中院驳回自然之友对国网甘肃公司弃风弃光诉讼：条件不符) (The Paper, 20 August 2018) <www.thepaper.cn/newsDetail_forward_2364374> accessed 20 September 2019.

91 See "Plaintiff's challenge", above.

92 FON's Update on the Gansu Grid Case (n 87); *FON* (北京市朝阳区自然之友环境研究所) *v. State Grid Gansu Electric Power Co.* (国网甘肃省电力公司) [2018] Gan Min Zhong No. 679 (Ruling, Higher People's Court of Gansu).

93 'Ruling' is a kind of judgment in China, which aims at procedural issues. See CPL 2017, art. 154.

the defendant can commit the alleged tortious act) backed by evidence is to be heard during an open court trial, unless the parties have reached a mediation agreement.[94] The second-instance court corrected the error and enabled the plaintiff to continue with a court trial to decide the substantive issue.

The Ningxia Grid Case has not seen any movement since acceptance, except for an informal "communication meeting" between the parties in April 2018.[95] In September 2019, FON informed the author that in neither of these cases has there been a pre-trial meeting, let alone a court trial.[96]

Although the reasons behind wind and solar power curtailment are complicated (e.g. the grid company cannot absorb all the renewable electricity),[97] and litigation may play a limited role in solving it, the NGO's proceeding at least makes an attempt at activating low-carbon mechanisms through litigation (in particular, the aforesaid REL mechanism for purchasing renewable electricity in the full amount). Moreover, given that NGOs are more willing than government authorities to share information on their cases' progress, through various platforms (e.g. websites, media, and social media accounts), the public-interest CCL filed by them opens a window for the public to see climate governance in China.

That having been said, the recent CCL practice is still restricted by the two challenges discussed above. First, since official data on emissions are difficult to collect and causation is nearly impossible to establish, the NGO plaintiff in the Gansu/Ningxia cases targeted grid companies, which have relatively clear duties under some low-carbon laws, instead of the real GHG emitters (i.e. coal-power plants). In other words, FON filed against the party that facilitated the emission of GHGs rather than the party that emitted them. Although these two kinds of conduct are both emission-related, the indirect strategy gives rise to legal controversy and slows the cases down (thus, Lanzhou Intermediate People's Court ruled against FON because it considered that the defendant was not responsible for the alleged damage). Although experimental litigation is indispensable for testing law enforcement, it will be more effective if it is supported by arrangements (e.g. GHG-emission standards) and law reform that assigns liability to entities that emit GHGs which exceed those standards.

Moreover, lacking guidelines on hearing this kind of lawsuit contributes to the time-consuming and pending situation of such cases. The Gansu Grid Case, first filed in 2016, was not accepted until mid 2017.[98] Normally, a court is to determine whether it accepts a case in seven days.[99] Requests for guidance are the main reason for delay. According to Wang Canfa, who is also the

94 CPL 2017, arts. 68, 133 and 134.

95 FON's 2018 Annual Report (n 3).

96 WeChat message from Liu Yina (an environmental lawyer of FON, who is managing these two cases) to author (12 September 2019). Pre-trial meeting is a formal process of Chinese civil procedure to exchange evidence and clarify disputes before a court trial. See CPL 2017, art.133.

97 Diao (n 84). See also Chapter 4 of this book.

98 FON's 2018 Annual Report (n 3).

99 CPL 2017, art.123.

head of the Center for Legal Assistance to Pollution Victims (one of the "supporting litigation units"[100] in this case), it took several months for the Lanzhou Intermediate People's Court to ask the SPC for instructions.[101] Even the acceptance process is tricky. At the time of writing, many months had passed since the second-instance court quashed the first-instance court's ruling and determined jurisdiction, yet no progress has been made.[102] Although information on the Ningxia Grid Case is limited, it can be inferred that similar issues contribute to its pending status, and that the court is being cautious and has not issued any judicial decision on it or held a pre-trial meeting to clarify the disputes. It is not realistic for the SPC to answer queries every time a local court accepts a CCL Core Case, but it is essential to guide the trial given the court's designated role in optimizing climate regulation.

Accordingly, there are two possible directions for improvement: first, implementation of the coordinated control of air pollutants and GHGs, mentioned earlier, which is a design element of the Law on Prevention and Control of Atmospheric Pollution, and, second, develop guidelines on tort-based GHG-emission CCL (which is important and feasible) for court reference. Because the method of legal transplant is not always convincing,[103] these optimizing directions will be based on the Chinese backdrop and practice. The following subsections will discuss these directions in more detail.

Carrying out the coordinated control of air pollutants and GHGs

Air pollution and climate change share common features. For instance, N_2O, which is regulated by the Chinese government as a kind of vehicle-engine pollutant,[104] is

100 According to CPL 2017 (art. 15), authorities, social groups, enterprises, and public institutions could support litigation. They are called supporting litigation units (支持起诉单位 or 支持起诉人 in Chinese) during the lawsuit. Another supporting litigation unit in this case is the Wind Energy Committee of the China Renewable Energy Society.

101 'Wang Canfa: The Practice of the EPIL Filed by Environmental NGOs and the Challenges It Faces' (王灿发：环保社会组织提起环境公益诉讼的实践与面临的挑战) (Center for Legal Assistance to Pollution Victims, 26 June 2017) <www.clapv.org/NewsContent.asp?id=1155&title=%D7%EE%D0%C2%B6%AF%CC%AC&titlecontent=NewsList&lei1=106> accessed 10 November 2019. Since the plaintiff's standing, defendant's information, claims, and jurisdiction were clear, the content of the first-instance court's query at that time was probably, "should we accept this new-type case involving a grid company's indirect role in air pollution?"

102 Normally, as per CPL 2017 (art. 149), a court must conclude a civil case in six months after it accepts it.

103 E.g. Pierre Legrand, 'The Impossibility of "Legal Transplants"' (1997) 4 *Maastricht Journal European & Comparative Law* 111, 120.

104 Announcement on Issuing National Emissions Standard 'Limits and Measurement Measures for Emissions from Light-duty Vehicles' (CHINA 6) (关于发布国家污染物排放标准《轻型汽车污染物排放限值及测量方法（中国第六阶段）》的公告), MEP Announcement No. (2016) 79, 23 December 2016. This standard will be applied to all to-be-sold and registered light-duty vehicles manufactured since 1 July 2020.

a GHG.[105] Nevertheless, air pollutants do not always play a role in climate change, so the remedy promoted by air-pollution EPIL cases would not always bring a decrease in GHG emissions if GHG and air-pollutant-control strategies are not coordinated.[106]

Central and local governments have already started to explore this mechanism. In October 2016, the State Council issued a work programme to curb GHG emissions, which emphasizes the coordinated control of carbon and air-pollutant emissions.[107] This document crystallizes not only GHG-intensity targets but also supporting arrangements, such as improving relevant rules and standards, strengthening GHG-emission statistics, and establishing a GHG-emission information-disclosure system (from the national level to the enterprise level). Under this instruction, authorities such as the MEP have taken some action. For example, the MEP developed a technology guideline in 2017 to help governmental environmental agencies calculate the amount of reduced pollutants and GHG-emission reductions in industrial enterprises,[108] which may encourage those enterprises to control the pollutant and GHG emissions at the same time if the related data is disclosed as a part of emission statistics. After all, the public could compare an enterprise's reduced emission amounts with those of another to evaluate their efforts. Moreover, according to Pkulaw.com (one of the biggest databases on Chinese laws and regulations), by August 2019, ten provincial-level administrative regions and 27 cities had promulgated their own regulations or guidelines to promote the coordinated control of GHGs and air pollutants.

As the earlier discussion shows, improving the GHG-emission information-disclosure system, relevant laws, and emission standards is crucial to the solution of CCL's difficulty in meeting its evidentiary burden. Carrying out the coordinated control of air pollutants and GHGs can be a direction for improvement. This co-governance mechanism introduces the successful experience of air-pollutant control into dealing with GHGs to decrease air pollutants and GHGs more effectively, which is also encouraged by the United Nations Environment Programme (UNEP).[109] However, the existing efforts focus mainly on the executive level rather than the judicial level. That is, the coordinated control currently relies on

105 See Annex A of the Kyoto Protocol to the United Nations Framework Convention on Climate Change.

106 Zhao, Lyu, and Wang (n 9), 375.

107 Notice of the State Council on Issuing the 13th Five-Year Work Program for Controlling GHG Emissions (国务院关于印发"十三五"控制温室气体排放工作方案的通知), State Council Issue No. (2016)61, 27 October 2016 (13th Five-Year Work Program for Controlling GHGs).

108 Notice of the General Office of the MEP on Issuing 'Technology Guideline on Calculating the Pollutant Removal and GHG Coordinated Control of Pollution Control Facilities of Industrial Enterprises' (环境保护部办公厅关于发布《工业企业污染治理设施污染物去除协同控制温室气体核算技术指南（试行）》的通知), MEP General Office Technology No. (2017) 73, 4 September 2017.

109 UNEP, 'Policies that Tackle Climate and Air Pollution at the Same Time Can Raise Global Climate Ambition' (UNEP, 18 October 2019) <www.unenvironment.org/news-and-stories/story/

the government, not the judiciary. As a result, the solution to the issue of proving damage and causation is left to Guiding Documents.

To remove the hurdles on proof, some of the aforesaid institutional designs on the coordinated control of air pollutants and GHGs should be further emphasized. First, the establishment of a GHG-emission information-disclosure system not only could provide emission evidence for CCL plaintiffs but could also benefit other low-carbon measures (e.g. the national carbon-emission trading market).[110] In the future, the Ministry of Ecology and Environment could integrate this information into the existing Administrative Information Platform of Pollutant Emission Permissions in China[111] for better public monitoring. Since March 2018, this ministry has taken over the government's duties to respond to climate change and reduce emissions.[112] Second, although GHG-emission standards for key enterprises (e.g. petrochemicals) are not realistic at present, because China is still peaking its carbon emissions and does not have a binding absolute GHG-emission-reduction target, some emission standards are still feasible, such as one for cars. The State Council's work programme on GHG-emission control requires research into a carbon-emission standard for new cars.[113] Third, when these supporting arrangements are mature, laws could be enacted or updated to prescribe the obligation to reduce GHG emissions for governments and enterprises and impose liability for exceeding corresponding standards. More-complete rules could help courts (e.g. by generating law-enforcement records for the court to reference) decide whether an amount of GHG emissions or a government's act has damaged environmental public interests. In other words, CCL Core Cases can play a more important role in China's climate regulation when the legal system is more complete.

Developing special guiding documents on tort-based GHG-Emission CCL

To improve the efficiency of hearings, since the existing practice is not sufficient to generate "Guiding Cases" or "Model Cases",[114] and other than by improving

policies-tackle-climate-and-air-pollution-same-time-can-raise-global-climate> accessed 5 December 2019.

110 E.g. IPE, 'Corporate Disclosure Crucial for National Carbon Market' (IPE, 13 December 2017) <wwwen.ipe.org.cn/reports/report_19476.html> accessed 26 September 2019.

111 '全国排污许可证管理信息平台' in Chinese. See <http://permit.mee.gov.cn/permitExt/syssb/xkgg/xkgg!licenseInformation.action>.

112 Decision of the First Session of the Thirteenth NPC on the State Council Institutional Reform Plan, Point 1.

113 13th Five-Year Work Program for Controlling GHGs, Point 4.

114 The SPC issues "Guiding Cases" (指导性案例) and "Model Cases" (典型案例) periodically to guarantee consistency in judicial decisions. See, e.g., Notice of the SPC on Issuing the 15th Batch of Guiding Cases (最高人民法院关于发布第15批指导性案例的通知), SPC No. (2016)449, 28 December 2016. The Ruitai Case mentioned above is a "Guiding Case" of this batch, so it is more influential than others. However, no "Guiding Case" or "Model Case" mentioning "climate change" or "GHGs" has been issued.

relevant laws, the SPC and other concerned authorities could consider developing special Guiding Documents to support tort-based GHG-emission CCL if more cases emerge. As previously mentioned, in theory, this kind of CCL includes mainly civil EPIL and private-interest-based tort litigation on GHG emissions. Such lawsuits would involve claims that the defendant assumes tort liability for its GHG-emission-related conduct, as well as damage related to this conduct; otherwise, they would be just normal EPIL or private-interest-based tort litigation. However, in practice, since it is nearly impossible to prove the causation between specific GHG emissions and individual damage, the following discussion will focus on civil EPIL on GHG emissions.

Key issues to be clarified are gaps in the verification of emission-related conduct, damage evaluation, causation identification, and remedy. Also, the coordination between related lawsuits is important to avoid conflicts.

For the verification of emission-related conduct, there is no doubt that the court can adopt the evidence from an authoritative GHG-emission information-disclosure system to confirm the defendant's emissions. If a local or national platform is not ready, governmental environmental agencies, which have the duty to deal with climate change now, could help the plaintiff establish the existence of the alleged emissions. The Hearing Guideline should include some arrangements (e.g. that the authority participates in the suit as a third party)[115] to involve the related agencies to assist the court in deciding facts, which is a demand of existing EPIL practice as well.[116] It also may help to investigate complicated emission-related issues involving governments, such as the wind and solar curtailment behind the FON's two grid cases.

Pursuant to the SPC's "judicial interpretation" on EPIL (Article 1), lawsuits against environmental pollution or ecological damage that has posed a "major risk" to public interests are also to be accepted by the courts if they comply with the aforesaid litigation requirements of CPL.[117] Nevertheless, this document does not specify the definition and features of "major risk". Some scholars argue that including "major risk" lowers the requirements for proving harm, from injury-in-fact to foreseeable harm,[118] but the lack of a clear guideline on identifying a "major risk" has brought more uncertainty to some preventative EPIL cases[119] which aim to protect endangered species (some of which are a carbon sink). This may also happen to tort-based GHG-emission CCL. Hence, improving the Hearing Guideline on verifying the "major risk" of harming environmental public interests, especially the "major risk" caused by GHG-emission-related conduct, is necessary as well. Moreover, plaintiffs usually need appraisal institutions to handle scientific issues, so encouraging the development of appraisal institutions

115 'Third party' is an existing mechanism of the CPL. See CPL 2017, art. 56.

116 See "Features of practice", above.

117 See "Plaintiff's challenge", above.

118 Zhao, Lyu, and Wang (n 9), 373.

119 E.g. the Yalung Case (accepted in December 2015) is still pending.

specializing in GHG-emission issues (including the verification of emissions, damage assessment, and causation analysis) would be of considerable benefit.

On damage evaluation, there should be more exploration on calculating the environmental damage caused by GHG emissions. If it is not possible to evaluate the environmental damage directly, could alternative measures be used to evaluate it? After all, in water pollution or air pollution, according to some MEP Appraisal Guidelines, in certain circumstances, the approximate environmental damage could be calculated on the basis of the disposal fee for excessive pollutants.[120] Similarly, environmental damage from an amount of GHG emissions could be based on the disposal fee of the excess emissions (if an emission standard is available). NGOs such as FON have held seminars on this issue.[121] The judiciary and administrative organs concerned should look into it as well. Other than appraisal institutions, governmental environmental agencies could also present an opinion on damage evaluation during the EPIL. As my prior empirical research (cited above) reveals, both of these professional forces could assist EPIL plaintiffs in calculating damage and ascertaining causation, so their role should be stressed further in a future Guiding Document of tort-based GHG-emission CCL.

On causation, as mentioned earlier, the burden of proof is, in a sense, favourable to the plaintiff,[122] and the winning rate of EPIL is high.[123] Nonetheless, according to a recent empirical study, most environmental tort plaintiffs still bear the initial liability to introduce evidence (usually in an official or certified report) to the judges when it comes to proving causation.[124] My own study confirms the court's reliance on appraisal reports from qualified experts or institutions. This phenomenon imposes an extra burden on the plaintiff, instead of shifting the burden. Therefore, a future Guiding Document not only needs to confirm that the favourable initial burden of proof is applicable to tort-based GHG-emission CCL, but should also clarify the requirements of establishing that the pollutant is related to the damage (e.g. through an expert opinion). Since "the environment" in an EPIL suit can be boundless,[125] under the light burden of proof it should be possible for a plaintiff of civil EPIL on GHG emissions to establish causation through proving the following elements: first, that the defendant emitted excessive GHGs (using official emission data and referencing a corresponding emission standard); second, that the amount of environmental damage (using alternative calculation methods) was caused by the defendant; and, third, that excessive GHG emissions are "related" to the environmental damage (by presenting an expert opinion).

120 E.g. Notice of the General Office of the MEP on Issuing the 'Recommended Methods for Environmental Damage Appraisal During the Emergency Disposal Stage of Environmental Emergency'.

121 FON's 2018 Annual Report (n 3).

122 See "Plaintiff's challenge", above.

123 See "Features of practice", above.

124 Fan Yang, Ting Zhang, and Hao Zhang, 'Adjudicating Environmental Tort Cases in China: Burden of Proof, Causation, and Insights from 513 Court Decisions' (2018) 21 *Asia Pacific Journal of Environmental Law* 171, 182, 188.

125 See "EPIL filed by NGOs", above.

Concerning remedies, because some concluded EPIL cases have positive results, the existing manifold remedies can continue to apply to the tort-based GHG-emission CCL.

Regarding the coordination between related lawsuits, there are two possible scenarios. The first is the coordination between civil EPIL on GHG emissions and general EPIL (e.g. air-pollution EPIL). If a company has polluted the environment (or damaged the ecology) and exceeded the GHG-emission standard during its production, a potential plaintiff could file only one EPIL lawsuit against it, because there is only one legal relationship: the defendant's production has harmed the public interests. If more than one plaintiff has filed EPIL against this conduct, the court should combine the hearings.[126] The second is the coordination between civil EPIL on GHG emissions filed by NGOs, civil EPIL on GHG emissions filed by procuratorates, and EEDCL on GHG emissions. Some "judicial interpretations" have already stipulated mechanisms for coordinating these three kinds of litigation systems.[127] Also, in light of the similarities between the EPIL and EEDCL, the arrangements for civil EPIL on GHG emissions in related guidelines could be applicable to EEDCL on GHG emissions.[128]

Conclusion: the breakthrough is just a start

Although CCL has become a global trend, the dominant scholarship has yet to notice the recent practice of China's CCL. FON's legal action against grid companies is said to be a breakthrough of CCL in this country. After careful examination, this chapter concludes that the EPIL filed by NGOs indeed brings an important development to China's CCL, given the clear climate-change-specific arguments of its recent cases, which did not appear before. In other words, it instigates the development of CCL Core Cases in China. Compared to its similar systems, this litigation system has started to deal with climate change issues directly. Its recent practice (i.e. the two grid cases) tries to implement low-carbon laws and control of air pollutants as well as GHGs from fossil fuels simultaneously.

Nonetheless, the EPIL filed by procuratorates cannot be ignored, as it is dominant in terms of caseload and can aim at local governments, corresponding to the first of the five trends in CCL mentioned by UNEP.[129] Administrative EPIL will become crucial if China's climate change regulation tightens and includes

126 Work Rules of the SPC on Hearing EPIL (For Trial Implementation), art. 17.

127 E.g. Judicial Interpretation on EEDCL, arts. 17 and 18.

128 E.g. the two "judicial interpretations" on civil EPIL and environmental tort can be applicable to EEDCL cases if there is no special provision. See ibid., art. 22.

129 The content of this trend is "holding governments to their legislative or policy commitments". The other four trends are (1) linking the impacts of resource extraction to climate change and resilience; (2) establishing that particular emissions are the proximate cause of particular adverse climate change impacts; (3) establishing liability for failure to adapt and the impacts of adaptation; (4) applying the public trust doctrine to climate change. See UNEP, *The Status of Climate Change Litigation: A Global Review,* 14 (UNON Publishing Services Section, 2017).

the government's duty to reduce GHG emissions in laws. In the future, China should emphasize the administrative EPIL on GHG emissions more, because, first, normally procuratorates have a higher capacity than NGOs and individuals; second, this kind of CCL Core Case would not face tricky hurdles in proving the damage and causation;[130] and third, it could facilitate the climate regulation more directly. Moreover, some CCL "Indirect" Cases filed by procuratorates (e.g. EPIL on ecological damage) have contributed to carbon-sink protection through urging defendants to remediate damaged vegetation (up to 1406.7 km^2 in 2018).[131] With regard to EEDCL, it also has potential, as more local governments give it a go, and the response to climate change becomes a part of the evaluation of local government performance.[132]

Instigating the development of CCL Core Cases is just the start, which is far from enough. The Gansu Grid Case and the Ningxia Grid Case are still waiting for their first court trial hearing and have not led to any obvious decarbonization effect. The pending situation reveals two challenges: a difficulty in proving GHG emissions, damage, and causation and a lack of Guiding Documents on hearing CCL, especially the emerging tort-based GHG-emission CCL. This chapter tries to point out feasible directions for improvement – carrying out the coordinated control of air pollutants and GHGs (mainly on the administrative level) and developing special Guiding Documents on tort-based GHG-emission CCL (mainly on the judicial level).

What will China's CCL Core Cases (currently tort-based GHG-emission CCL cases) be like in the future? Scholars such as Zhao Yue contend that the future scope for tort-based CCL in China will hinge on how strict climate change regulation becomes and how proactive authorities are in its enforcement.[133] According to the previous analysis, this is reasonable to a great extent, because China's climate regulation is currently government led and because its CCL Core Cases rely on the establishment of relevant supporting arrangements, such as the GHG-emission information-disclosure system. Although the attempts of NGO pioneers cannot be ignored, there are too many uncertainties surrounding the CCL Core Cases in China: the official GHG-emission information is limited; there is no effective emission standard to reference; no appraisal institution is able to evaluate causation; the court does not have a guideline to streamline its hearings; and

130 According to the Administrative Litigation Law and its Guiding Documents, to establish a relevant authority's liability, a procuratorate plaintiff needs to prove only: (1) that the defendant has a duty to reduce GHG emissions according to the law; (2) that the defendant is not performing this duty properly and damages the public interests (for example, the plaintiff could compare the actual GHG-emission intensity of the region with the corresponding GHG intensity targets which, for certain provinces, have already been disclosed by documents like the 13th Five-Year Work Program for Controlling GHGs); and that the plaintiff has sent a "Procuratorial Recommendation" to the defendant during the pre-trial process to urge it to correct its nonfeasance or misfeasance, but it has not done so.

131 2018 Annual Report of the SPP (n 15).

132 13th Five-Year Work Program for Controlling GHGs.

133 Zhao, Lyu, and Wang (n 9), 376.

so on. It can be expected that the Chinese CCL Core Cases will develop only slowly in the near future given the deficiencies in supporting mechanisms, which in essence reflects the fact that China does not have a binding absolute GHG-reduction target, and most emission-reduction mechanisms have been only newly launched.[134] The struggle of the two recent cases embodies this challenging exploration. Other NGOs and eligible parties may follow and initiate relatively indirect or simple CCL cases (e.g. lawsuits similar to the grid cases or tort-based CCL cases on carbon-sink protection), but such symbolic efforts may play a more important role in raising public awareness than in actual decarbonization. At present, civil society's push is not sufficient to address climate-change-related issues directly through litigation. As for courts, their activism plays an important role in the EPIL filed by NGOs, so the judiciary's attitude also matters. Needless to say, developing the SPC's Guiding Documents is a key direction for improvement. Therefore, the prospect for China's CCL Core Cases depends mainly on the advancement and interaction of governments, courts, and EPIL plaintiffs (especially NGOs that act as pioneers due to their flexibility).

134 e.g. The 13th Five-Year Work Program for Controlling GHGs, Point 1.

Bibliography

A. International Treaties and Decisions

Kyoto Protocol, Decision 3/CMP.1, Modalities and Procedures for a Clean Development Mechanism as defined in Article 12 of the Kyoto Protocol, Conference of the Parties Serving as the Meeting of the Parties to the Kyoto Protocol, Report of the Conference of the Parties serving as the meeting of the Parties to the Kyoto Protocol on its First Session, held at Montreal from 28 November to 10 December 2005 – Addendum–Part 2: Action Taken by the Conference of the Parties Serving as the Meeting of the Parties to the Kyoto Protocol, UN Doc FCCC/KP/CMP/2005/5/Add.1 (30 March 2006).

Kyoto Protocol, Decision 15/CMP.1, Guidelines for the Preparation of the Information Required under Article 7 of the Kyoto Protocol FCCC/KP/CMP/2005/8/Add.2 (30 November 2005).

Kyoto Protocol, Decision 19/CMP.1, Guidelines for National Systems under Article 5, Paragraph 1, of the Kyoto Protocol FCCC/KP/CMP/2005/8/Add.3 (30 November 2005).

Kyoto Protocol, Decision 22/CMP.1, Guidelines for Review under Article 8 of the Kyoto Protocol FCCC/KP/CMP/2005/8/Add.3 (2005).

Kyoto Protocol to the United Nations Framework Convention on Climate Change, UN Doc FCCC/CP/1997/7/Add 1 (10 December 1997) (entered into force 16 February 2005).

Paris Agreement, Decision 4/CMA.1, Further Guidance in Relation to the Mitigation Section of Decision 1/CP.21, FCCC/PA/CMA/2018/3/Add.1 (2018).

Paris Agreement, Decision 18/CMA.1, Modalities, Procedures and Guidelines for the Transparency Framework for Action and Support Referred to in Article 13 of the Paris Agreement, FCCC/PA/CMA/2018/3/Add.2 (2018).

UNFCCC, Decision 1/CP.13, Bali Action Plan FCCC/CP/2007/6/Add.1 (2007).

UNFCCC, Decision 1/CP.16, The Cancun Agreements: Outcome of the Work of the Ad Hoc Working Group on Long-Term Cooperative Action under the Convention FCCC/CP/2010/7/Add.1 (2010).

UNFCCC, Decision 1/CP.21, Adoption of the Paris Agreement, FCCC/CP/2015/10/Add.1 (2015).

UNFCCC, Decision 2/CP.1, Review of First Communications from the Parties Included in Annex I to the Convention FCCC/CP/1995/7/Add.1 (1995).

UNFCCC, Decision 2/CP.15, Copenhagen Accord FCCC/CP/2009/11/Add.1 (2009).

UNFCCC, Decision 2/CP.17, Outcome of the Work of the Ad Hoc Working Group on Long-Term Cooperative Action under the Convention FCCC/CP/2011/9/Add.1 (2011).

UNFCCC, Decision 4/CP.5, Guidelines for the Preparation of National Communications by Parties Included in Annex I to the Convention, Part II: UNFCCC Reporting Guidelines on National Communications FCCC/CP/1999/6/Add.1 (1999).

UNFCCC, Decision 4/CP.15, Methodological Guidance for Activities Relating to Reducing Emissions from Deforestation and Forest Degradation FCCC/CP/2009/11/Add.1 (2009).

UNFCCC, Decision 10/CP.13, Compilation and Synthesis of Fourth National Communications FCCC/CP/2007/6/Add.1 (2007).

UNFCCC, Decision 11/CP.4, National Communications from Parties Included in Annex I to the Convention FCCC/CP/1998/16/Add.1 (1998).

UNFCCC, Decision 17/CP.7, Report of the Conference of the Parties on Its Seventh Session, Held at Marrakesh from 29 October to 10 November 2001 – Addendum–Part 2: Action Taken by the Conference of the Parties, UN Doc FCCC/CP/2001/13/Add.1/Corr.1 (15 May 2002).

UNFCCC, Decision 17/CP.8, Guidelines for the Preparation of National Communications from Parties Not Included in Annex I to the Convention FCCC/CP/2002/7/Add.2 (2003).

UNFCCC, Decision 18/CP.8, Guidelines for the Preparation of National Communications by Parties Included in Annex I to the Convention, Part I: UNFCCC Reporting Guidelines on Annual Inventories FCCC/CP/2002/7/Add.2 (2002).

UNFCCC, Decision 19/CP.8, Guidelines for the Technical Review of Greenhouse Gas Inventories from Parties Included in Annex I to the Convention FCCC/CP/2002/7/Add.2 (2002).

UNFCCC, Decision 20/CP.19, Composition, Modalities and Procedures of the Team of Technical Experts under International Consultation and Analysis FCCC/CP/2013/10/Add.2 (2013).

UNFCCC, Decision 24/CP.19, Revision of the UNFCCC Reporting Guidelines on Annual Inventories for Parties Included in Annex I to the Convention FCCC/CP/2013/10/Add.3 (2013).

UNFCCC, Guidelines on Reporting and Review FCCC/CP/1999/7 (1999).

UNFCCC, Guidelines on Reporting and Review FCCC/CP/2002/8 (2002).

UNFCCC, Report of the Conference of the Parties on its Thirteenth Session, held in Bali from 3 to 15 December 2007 FCCC/CP/2007/6/Add.1 (14 March 2008).

UNFCCC, Subsidiary Body for Scientific and Technical Advice, Guidelines for the Preparation of National Communications by Parties Included in Annex I to the Convention, Part I: UNFCCC Reporting Guidelines on Annual Inventories (Following Incorporation of the Provisions of Decision 13/CP.9) FCCC/SBSTA/2004/8 (2004).

UNFCCC, Subsidiary Body for Scientific and Technical Advice, Updated UNFCCC Reporting Guidelines on Annual Inventories Following Incorporation of the Provisions of Decision 14/CP.11 FCCC/SBSTA/2006/9 (2006).

UNFCCC (opened for signature 9 May 1992, entered into force 21 March 1994), 1771 U.N.T.S. 107.

B. Laws and Regulations

Administrative Measures for the Operation and Management of Clean Development Mechanism Projects (清洁发展机制项目运行管理办法) National Development and Reform Commission, Ministry of Science and Technology, Ministry of Foreign Affairs, and Ministry of Finance; adopted in 2005, amended by Order No. (2011) 11, 3 August 2011.

Agriculture Law of the People's Republic of China (中华人民共和国农业法) Standing Committee of the National People's Congress (adopted on 2 July 1993).

Circular Economy Promotion Law of the People's Republic of China (中华人民共和国循环经济促进法) Standing Committee of the National People's Congress (adopted on 29 August 2008, amended on 26 October 2018).

Cleaner Production Promotion Law of the People's Republic of China (中华人民共和国清洁生产促进法) Standing Committee of the National People's Congress (adopted on 29 June 2002, amended on 29 February 2012).

Coal Industry Law of the People's Republic of China (中华人民共和国煤炭法) Standing Committee of the National People's Congress (adopted on 29 August 1996, amended on 27 August 2009, 22 April 2011, 29 June 2013, and 7 November 2016).

Commission Regulation (EU) No 550/2011 of 7 June 2011 on Determining, Pursuant to Directive 2003/87/EC of the European Parliament and of the Council, Certain Restrictions Applicable to the Use of International Credits from Projects Involving Industrial Gases [2011] OJ L149/1.

Constitution of the People's Republic of China (中华人民共和国宪法) National People's Congress (adopted on 4 December 1982, amended on 12 April 1988, 29 March 1993, 15 March 1999, 14 March 2004, and 11 March 2018).

Directive 2003/87/EC of the European Parliament and of the Council of 13 October 2003 Establishing a Scheme for Greenhouse Gas Emission Allowance Trading within the Community and Amending Council Directive 96/61/EC, OJ L 275, 25 October 2003.

Directive 2004/101/EC of the European Parliament and of the Council of 27 October 2004 amending Directive 2003/87/EC Establishing a Scheme for Greenhouse Gas Emission Allowance Trading within the Community, in respect of the Kyoto Protocol's Project Mechanisms [2004] OJ L338/18.

Electric Power Law of the People's Republic of China (中华人民共和国电力法) Standing Committee of the National People's Congress (adopted on 28 December 1995, amended on 27 August 2009, 24 April 2015, and 29 December 2018).

Energy Conservation Law of the People's Republic of China (中华人民共和国节约能源法) Standing Committee of the National People's Congress (adopted on 1 November 1997, amended on 2 July 2016 and 26 October 2018).

Environmental Impact Assessment Law (环境影响评价法) Standing Committee of the National People's Congress (adopted on 28 October 2002, amended on 2 June 2016 and 29 December 2018).

Environmental Protection Law of the People's Republic of China (中华人民共和国环境保护法) Standing Committee of the National People's Congress (adopted on 26 December 1989, amended on 24 April 2014).

Environmental Protection Tax Law of the People's Republic of China (2018 Revision) (中华人民共和国环境保护税法(2018修正)) Standing Committee of the National People's Congress, 26 October 2018.

Forest Law of the People's Republic of China (中华人民共和国森林法) Standing Committee of the National People's Congress (adopted on 20 September 1984, amended on 27 August 2009).

General Principles of the Civil Law of the People's Republic of China (中华人民共和国民法通则) National People's Congress (adopted on 12 April 1986, amended on 27 August 2009).

Grassland Law of the People's Republic of China (中华人民共和国草原法) Standing Committee of the National People's Congress (adopted on 18 June 1985, amended on 27 August 2009 and 29 June 2013).

Interim Measures for the Administration of Voluntary Greenhouse Gas Emission Reduction Transactions (温室气体自愿减排交易管理暂行办法) NDRC Order No. (2012)2862, 9 October 2012.

Interim Measures on Managing the Trading of Financial Derivatives Developed by Financial Institutions (金融机构衍生产品交易业务管理暂行办法), China Banking Regulatory Commission (adopted on 28 December 2006 and amended on 27 January 2011).

Interim Regulations on the Management of Carbon Emissions Trading (Draft for Comment) (碳排放权交易管理暂行条例（征求意见稿）), Ministry of Ecology and Environment, 29 March 2019.

Law on Combating Climate Change of the People's Republic of China (Draft for Public Comments) (中华人民共和国气候变化应对法（征求意见稿）) (18 March 2012) <http://news.china.com.cn/txt/2012-03/18/content_24923504.htm>.

Law on Prevention and Control of Atmospheric Pollution (大气污染防治法), Standing Committee of the National People's Congress (adopted on 5 September 1987, amended on 29 August 1995, 29 April 2000, 29 August 2015, and 26 October 2018).

Legislation Law of the People's Republic of China (中华人民共和国立法法) National People's Congress (adopted on 15 March 2000, amended on 15 March 2015).

Measures for the Operation and Management of the Clean Development Mechanism Projects (清洁发展机制项目运行管理办法) NDRC, Ministry of Science and Technology, Ministry of Foreign Affairs and Ministry of Finance Order No. (2011)11, 3 August 2011.

Meteorology Law of the People's Republic of China (中华人民共和国气象法) Standing Committee of the National People's Congress (adopted on 31 October 1999, amended on 27 August 2009, 31 August 2014, and 7 November 2016).

Qinghai Provincial Government, Measures of Responding to Climate Change in Qinghai Province (青海省应对气候变化办法), Qinghai Provincial Government Order No. (2010)75, 23 July 2010.

Regulation of Environmental Management of Construction Projects (建设项目环境保护管理条例) State Council Order No. (2017)682, 1 October 2017.

Regulation on Energy Conservation for Civil Buildings (民用建筑节能条例) State Council Order No. (2008)530, 23 July 2008.

Regulation on Energy Conservation Management for Public Buildings (公共机构节能条例) State Council Order No. (2008) 531, 1 August 2008, amended on 1 March 2017.

Regulations on Managing the Emissions Trading Scheme (Review Draft) (碳排放权交易管理条例(送审稿)), NDRC No. (2016), 2016.

Renewable Energy Law of the People's Republic of China (中华人民共和国可再生能源法) Standing Committee of the National People's Congress (adopted on 28 February 2005, amended on 26 December 2009).

Shanxi Provincial Government, Measures on Responding to Climate Change in Shanxi Province (山西省应对气候变化办法), Shanxi Provincial Government Order No. (2011)19, 23 August 2011.

Water and Soil Conservation Law of the People's Republic of China (中华人民共和国水土保持法) Standing Committee of the National People's Congress (adopted on 29 June 1991, amended on 25 December 2010).

Water Law of the People's Republic of China (中华人民共和国水法) Standing Committee of the National People's Congress (adopted on 21 January 1988).

Water Pollution Prevention and Control Law (水污染防治法), Standing Committee of the National People's Congress (adopted on 11 May 1984, amended on 15 May 1996, 28 February 2008, and 27 June 2017).

C. Chinese Government Policy and Information Documents

Allocation Plan for the Allowances under Hubei's Emissions Trading Pilot Program (湖北省碳排放权配额分配方案) Hubei Development and Reform Commission, 22 May 2014.

Allocation Plan for the Allowances under Hubei's Emissions Trading Pilot Program in 2017 (湖北省2017年碳排放权配额分配方案) Hubei Development and Reform Commission, 10 January 2018.

China's National Climate Change Programme (中国应对气候变化国家方案) State Council Order No. (2007)17, 3 June 2007.

China's Policies and Actions for Addressing Climate Change (2018) (Ministry of Ecology and Environment, November 2018) <http://english.mee.gov.cn/News_service/news_release/201812/P020181203536441502157.pdf>.

China's Policies and Actions on Climate Change (2011) (中国应对气候变化的政策与行动 (2011)), 'Information Office of the State Council of the People's Republic of China' <www.china.com.cn/ch-book/node_7142682.htm>.

Chinese Academy of Social Sciences, Statement on the Law on Addressing Climate Change (Proposal) (关于《气候变化应对法》(建议稿)的说明), 2012.

Department of Climate Change, 'Enhanced Actions on Climate Change: China's Intended Nationally Determined Contributions' (NDRC, 2015).

Energy Research Institute, NDRC, 'Research Report on China's Solar PV Industry Development' (中国光伏产业发展研究报告) (August 2006) <http://zsdh.library.sh.cn:8080/FCKeditor/filemanager/upload/jsp/UserImages/1216361096343.pdf>.

Fourth Group of National Coordination Committee on Climate Change (国家气候变化协调小组第四工作组), 'China's Preliminary Draft of International Convention on Climate Change' (《关于气候变化的国际公约》条款草案（中国的建议）), in Secretariat of Environmental Protection Commission of the State Council (eds) (国务院环境保护委员会秘书处编), *Corpus of Documents of Environmental Protection Commission of the State Council* (国务院环境保护委员会文件汇编) (Volume 2, China Environmental Science Press, 1995).

Government of China, Second Biennial Update Report on Climate Change (Beijing, State Council, 2018).

General Guideline on GHG Emissions Accounting and Reporting for Industrial Enterprises (工业企业温室气体排放核算和报告通则), AQSIQ and SAC GB/T32150–2015 (effective on 1 June 2016).

General Specifications on GHG Emissions Verification for Industrial Enterprises (行业企业温室气体排放核查通用规范), CNCA RB/T211–2016, June 2017.

Guidelines of the Eleventh Five-Year Plan for National Economic and Social Development (中华人民共和国经济和社会发展第十一个五年计划), March 2006.

Guidelines on Validation and Verification for Voluntary Greenhouse Gas Emission Reduction Projects (温室气体自愿减排项目审定与核证指南) NDRC Order No. (2012)2862, 9 October 2012.

Implementation Measures for Allocating and Managing Emission Allowances in Shanghai (上海市2013–2015碳排放配额分配和管理方案) Shanghai Development and Reform Commission Order No. (2013)168, 22 November 2013.

Implementation Measures of Hubei's Emissions Trading Pilot Program (湖北省碳排放权交易试点工作实施方案), Hubei People's Government Order No. (2013)9, 18 February 2013.

Implementation Measures to Resolve the Curtailment of Hydro, Wind and Solar Energy (解决弃水弃风弃光问题实施方案) NDRC and NEA Order No. (2017)1942, 8 November 2017.

Implementation Opinions of the State Council on Rectification of Various Types of Trading Platforms (国务院办公厅关于清理整顿各类交易场所的实施意见), State Council Order No. (2012)37, 12 July 2012.

Implementation Plan for Allowances Allocation Guangdong in 2018 (广东省2018年度碳排放配额分配实施方案), Guangdong Development and Reform Commission, 26 July 2018.

Information Office of the State Council, China's Policies and Actions for Addressing Climate Change, 2008 (中国应对气候变化的政策与行动 (2008)), October 2008.

Information Office of the State Council, China's Policies and Actions for Addressing Climate Change, 2011 (中国应对气候变化的政策与行动 (2011)), November 2011.

Information Office of the State Council, China's Policies and Actions for Addressing Climate Change, 2018 (中国应对气候变化的政策与行动 (2018)), October 2018.

Interim Measures for Managing and Trading Emissions Allowances in Guangdong Province (广东省碳排放管理试行办法), Guangdong People's Government Order No. (2014)197, 15 January 2014.

Interim Measures for Managing and Trading Emissions Allowances in Hubei Province (湖北碳排放权管理和交易暂行办法), Hubei People's Government Order No. (2014)371, 23 April 2014.

Interim Measures for the Administration of Carbon Emission Permit Trading (碳排放权交易管理暂行办法), NDRC Order No. (2014)17, 10 December 2014.

Interim Measures for the Administration of Carbon Emissions Trading in Shenzhen (深圳市碳排放权交易管理暂行办法), Shenzhen People's Government Order No. (2014)262, 19 March 2014.

Interim Rules of Trading Carbon Emission Allowances at Beijing Environmental Exchange (北京环境交易所碳排放权交易规则 (试行)) <http://images.bjets.com.cn/www/201501/20150113171115897.pdf>.

Measure for Renewable Electricity Quota and Assessment (Draft for Comment) (可再生能源电力配额及考核办法(征求意见稿)), National Energy Administration Order No. (2018)28, 23 March 2018.

Measures of Accountability Assessment with Regard to Fulfilment of the Control Targets Related to CO_2 Emissions per Unit of GDP (单位国内生产总值二氧化碳排放降低目标责任考核评估办法), NDRC Order No. (2014)1828, August 2014.

Ministry of Ecology and Environment, Third Communication on Climate Change (中华人民共和国气候变化第三次国家信息通报), 2018.

National Energy Commission, 'Coalmine Approvals Suspended in the Next Three Years' (国家能源局：未来三年暂停煤矿审批), 30 December 2015, <http://coalchem.anychem.com/2015/12/30-16899.html>.

National People's Congress, Thirteenth Five-Year Plan for National Economic and Social Development (国民经济和社会发展第十三个五年规划), 17 March 2016.

National People's Congress, Twelfth Five-Year Plan for National Economic and Social Development (国民经济和社会发展第十二个五年规划), 13 March 2011.

National Strategy of Climate Change Adaptation (国家适应气候变化战略) <www.gov.cn/gzdt/att/att/site1/20131209/001e3741a2cc140f6a8701.pdf>.

NDRC, China Medium and Long Term Energy Conservation Plan (25 November 2004) <fourfact.com/images/uploads/China_Energy_Saving_Plan.pdf>.

NDRC, China's Policies and Actions for Addressing Climate Change (Beijing, October 2017).

NDRC, City Action Plan of Climate Change Adaptation (城市适应气候变化行动方案), NRDC Order No. (2016)245.

NDRC, 'Enhanced Actions on Climate Change' <http://www4.unfccc.int/ndcregistry/PublishedDocuments/China%20First/China%27s%20First%20NDC% 20Submission.pdf>.

NDRC, National Plan to Address Climate Change (2014–20) (国家应对气候变化规划(2014–20)), NRDC Order No. (2014)2347, 17 September 2014.

NDRC, Notice about Distributing the Pilot Work of Building Climate-resilient Cities (关于印发气候适应型城市建设试点工作的通知), NRDC Order No. (2017)343, 21 February 2017.

NEA, '2016 Northwest Renewable Energy Integration Report' (2016年西北区域新能源并网ff行情况通报) (19 January 2017) <www.nea.gov.cn/2017-01/19/c_135996630.htm>.

NEA, 'Operational Condition of Wind Power Connecting to Grid in 2016' (2016年风电并网运行情况) (NEA, 26 January 2017) <www.nea.gov.cn/201701/26/c_136014615.htm>.

Notice of MOF and MOHURD on Solar Roofs Program (财政部和住房和城乡建设部印发关于实施太阳能屋顶计划的通知), MOF and MOHURD Order No. (2009)34, 16 April 2009.

Notice of MOF on Provisional Administrative Measures on the Renewable Energy Development Fund (财政部关于印发可再生能源发展专项资金管理暂行办法的通知), MOF Order No. (2006)237, 30 May 2006.

Notice of MOF, MOST, and NEA on Golden Sun Demonstration Project (财政部，科技部和国家能源局关于印发金太阳示范工程财政补助资金管理暂行办法的通知), MOF, MOST and NEA, Order No. (2009)397, 21 July 2009.

Notice of NDRC and NEA on Establishing and Improving the Safeguard Mechanism for Renewable Energy Consumption (国家发展改革委国家能源局关于建立健全可再生能源电力消纳保障机制的通知), NDRC No. (2019)807, 10 May 2019.

Notice of NDRC on Abolishing the Requirement on the Rate of Localization of Equipment Procurement on Wind Power Projects (国家发改委下发国家发展改革委员会关于取消风电工程项目设备采购国产化率要求的通知), NDRC Order No. (2009)2991, 25 November 2009.

Notice of NDRC on Guiding Catalogue for Development of the Renewable Energy Industry (国家发展改革委关于印发可再生能源产业发展指导目录的通知), NDRC Order No. (2005)2517, 29 November 2005.

Notice of NDRC on Initiating the Compilation of Provincial Greenhouse Gas Inventory (关于启动省级温室气体清单编制工作有关事项的通知), NDRC Order No. (2010)2350, September 2010.

Notice of NDRC on Issuing Provisional Administrative Measures on Pricing and Cost Sharing for Renewable Energy Power Generation (国家发展改革委关于印发可再生能源发电价格和费用分摊管理试行办法的通知), NDRC Order No. (2006)7, 20 January 2006.

Notice of NDRC on Issuing Provisional Regulation on Renewable Energy Surcharge Balancing (国家发展改革委关于印发可再生能源电价附加收入调配暂行办法), NDRC Order No. (2007)44, 11 January 2007.

Notice of NDRC on Regulation on the Administration of Power Generation from Renewable Energy (国家发展改革委关于印发可再生能源发电有关管理规定), NDRC Order No. (2006)13, 5 January 2006.

Notice of NEA on Establishing a Surveillance and Early Warning Mechanism for Promoting Sustainable and Healthy Development of the Wind Power Industry (国家能源局关于建立监测预警机制促进风电产业持续健康发展的通知), NEA Order No. (2016)196, 18 July 2016.

Notice of NEA on Promoting the Top-Runner Program and Requirements of Implementing the Program in 2017 (国家能源局关于推进光伏发电领跑者计划实施和2017年领跑基地建设有关要求的通知), NEA Order No. (2017)54, 22 September 2017.

Notice of the General Office of Hubei Province on the Adjustment of the Name and Members of the Leading Group on Energy Conservation (Climate Change) (省人民政府办公厅关于湖北省节能减排（应对气候变化）工作领导小组更名并调整组成人员的通知) General Office of Hubei Province Order No. (2019)22, 20 March 2019.

Notice of the General Office of the State Council Regarding the Changes of Members of the National Leading Group to Address Climate Change and Energy Conservation (国务院办公厅关于调整国家应对气候变化及节能减排工作领导小组组成人员的通知(2019)) General Office of the State Council Order No. (2019)99, 7 October 2019.

Notice of the Hubei DRC on Issuing the Allocation Plan for the Allowances under Hubei's Emissions Trading Pilot Program in 2015 (省发展改革委关于印发《湖北省2015年碳排放权配额分配方案》的通知), Hubei DRC Order No. (2015)708, 19 November 2015.

Notice of the Hubei DRC on Issuing the Allocation Plan for the Allowances under Hubei's Emissions Trading Pilot Program in 2016 (省发展改革委关于印发《湖北省2016年碳排放权配额分配方案》的通知), Hubei DRC Order No. (2016)791, 30 December 2016.

Notice of the NDRC on Initiating Pilot Emissions Trading Programs (国家发展改革委办公厅关于开展碳排放权交易试点工作的通知) NDRC Order No. (2011)2601, 29 October 2011.

Notice of the NDRC on Issuing the Plan for Constructing the National Emissions Trading Scheme (Electricity Generation Sector) (国家发展改革委关于印发《全国碳排放权交易市场建设方案（发电行业）》的通知), NDRC No. (2017)2191, 18 December 2017.

Notice of the Shenzhen DRC on Implementing the Carbon Emissions Trading Program in 2016 (深圳市发展和改革委员会关于开展2016年度碳排放权交易工作的通知), Shenzhen DRC, 18 September 2016.

Notice of the State Council on the Establishment of Institutions of the Central Government (国务院关于机构设置的通知), State Council No. (2018)6, 22 March 2018.

Notice of the State Council on Issuing the Work Plan to Control Greenhouse Gas Emissions during the 13th Five-Year Period (国务院关于印发"十三五"控制温室气体排放工作方案的通知), State Council Order No. (2016)61, 27 October 2016.

Notice of the State Council Regarding the Establishment of the National Leading Group to Address Climate Change and Energy Conservation (国务院关于成立国家应对气候变化及节能减排工作领导小组的通知) State Council Order No. (2007)18, 12 June 2007.

Notice on Compiling the List of the Liable Entities in the Power Sector for the National ETS and Surrendering Relevant Materials (关于做好全国碳排放权交易市场发电行业重点排放单位名单和相关材料报送工作的通知), MEE Order No. (2019)528, 27 May 2019.

Notice on Establishing Contacts between the Sellers and Buyers with the Intention to Transact CCERs (核证自愿减排量意向交易方征集活动的通知) (Beijing Environmental Exchange, 2014) <www.bjets.com.cn/article/xxzx/jysgg/201403/20140300000159.shtml>.

Notice on Issuing GHG Emission Accounting Methods and Reporting Guidelines for Enterprises of the First Batch (Ten Industries Involved) (Trial) (关于印发首批10 个行业企业温室气体排放核算方法与报告指南试行的通知), NDRC Order No. (2013)2526, October 2013.

Notice on Issuing GHG Emission Accounting Methods and Reporting Guidelines for Enterprises of the Second Batch (Four Industries Involved) (Trial) (关于印发第二批4个行业企业温室气体排放核算方法与报告指南（试行）的通知), NDRC Order No. (2014)2920, December 2014.

Notice on Issuing GHG Emission Accounting Methods and Reporting Guidelines for Enterprises of the Third Batch (Ten Industries Involved) (Trial) (关于印发第三批10个行业企业温室气体排放核算方法与报告指南（试行）的通知), NDRC Order No. (2015)1722, July 2015.

Notice on Launching Climate Change Statistical Operations (关于开展应对气候变化统计工作的通知), NDRC and NBS Order No. (2013)80, November 2013.

Notice on Releasing the Guidelines for Preparation of Provincial GHG Inventory (Trial) (关于印发省级温室气体清单编制指南（试行）的通知), NDRC Order No. (2011)1041, March 2011.

Notice on the Preparation of GHG Emission Reports by Key Enterprises (Institutions) (关于组织开展重点企（事）业单位温室气体排放报告工作的通知), NDRC Order No. (2014)63, January 2014.

Opinions of the Supreme People's Court on Giving Full Play to Judicial Functions to Provide Judicial Services and Guarantee for Promoting Ecological Civilization Construction and Green Development (最高人民法院关于充分发挥审判职能作用为推进生态文明建设与绿色发展提供司法服务和保障的意见) Supreme People's Court Order No. (2016)12, 2 June 2016.

Opinions on Strengthening Statistical Work related to Combating Climate Change (关于加强应对气候变化统计工作的意见), NDRC and NBS Order No. (2013)937, October 2013.

Progress of China's Carbon Market 2017 (2017中国碳市场进展) (Energy Research Institute of NDRC and Environmental Defense Fund Report, 2017) <www.cet.net.cn/uploads/soft/201805/1_31184021.pdf>.

Provisions of the Supreme People's Court Concerning Work on Case Guidance (关于案例指导工作的规定), Supreme People's Court Order No. (2010)51, 26 November 2010.

Requirements of Greenhouse Gas Emission Accounting and Reporting: Part 1 to Part 10 (温室气体排放核算与报告要求：第1部分 至 第10部分), AQSIQ and SAC GB/T32151.1–10–2015 (effective on 1 June 2016).

Resolution of the Standing Committee of the National People's Congress on Making Active Responses to Climate Change (全国人民代表大会常务委员会关于积极应对气候变化的决议), Standing Committee of the National People's Congress, 27 August 2009.

SCIO Briefing on China's Policies and Actions for Addressing Climate Change <http://english.scio.gov.cn/pressroom/2019-11/28/content_75458089.htm>.

Shanghai Clearing House, 'Introduction to the Emission Allowances Futures in Shanghai' (上海碳配额远期产品介绍) (15 December 2016) <www.shclearing.com/sheaf/ywjs_sheaf/cpjs/201612/t20161215_211339.html>.

Speech at the High Level Segment of COP16 and CMP6 Delivered by Vice-Chairman Xie Zhenhua, National Reform and Development Commission, People's Republic of China, Cancun (8 December 2010).

Speech by Liu Jiang, Head of China's Delegation to the 5th Conference of the Parties to the UNFCCC in 1999 (18 July 2002).

Speech by Liu Jiang, Head of China's Delegation to the 7th Conference of the Parties to the UNFCCC in 2001 (18 July 2002).

Speech by Xie Zhenhua, Head of China's Delegation at the 13th Conference of the Parties to the UNFCCC (12 December 2007).

State Council of the People's Republic of China, Energy Development Strategic Action Plan (2014–2020) (能源发展战略行动计划（2014–2020年）) <www.gov.cn/zhengce/content/2014-11/19/content_9222.htm>.

State Council Information Office, China, 'China's Efforts and Achievements in Promoting the Rule of Law' (2008) 7 *Chinese Journal of International Law* 513.

State Council Institutional Reform Plan of 2018 (2018年国务院机构改革方案) National People's Congress, effective as of 17 March 2018.

State Council, Second National Communication on Climate Change of China (气候变化第二次国家信息通报), 2012.

Technical Guidelines for Environmental Impact Assessment (General Principles), Ministry of Environmental Protection (规划环境影响评价技术导则(总纲) HJ 130–2014, 2014.

Understanding the Initiation of the National ETS by Looking at the Suspension of the CCERs (从暂缓CCER看全国碳市场启动) (China Economic Herald, managed by the NDRC, 7 April 2017) <www.ceh.com.cn/ztbd/jnjpzk/1029753.shtml>.

Work Plan to Control Greenhouse Gas Emissions during the 12th Five-Year Period ("十二五"控制温室气体排放工作方案), State Council Order No. (2011)41, 1 December 2011.

Work Scheme for Statistical Operations Related to Addressing Climate Change (应对气候变化统计工作方案), NBS Order No. (2014)7, January 2014.

D. Scholarly and Other Works

Adger, Neil et al., 'Moral Reasoning in Adaptation to Climate Change' (2017) 26(3) *Environmental Politics* 371.

Andrews-Speed, Philip, and Sufang Zhang, *China as a Global Clean Energy Champion: Lifting the Veil* (Springer, 2019).

Arup, Christopher, and Hao Zhang, 'Lessons from Regulating Carbon Offset Markets' (2015) 4(1) *Transnational Environmental Law* 69.

Averchenkova, Alina, 'Legislating for a Low Carbon and Climate Resilient Transition: Learning from International Experiences' (Real Instituto Elcano, Elcano Policy Paper, January 2019).

Averchenkova, Alina et al., 'Climate Policy in China, the European Union and the United States: Main Drivers and Prospects for the Future' (Grantham Research Institute on Climate Change and the Environment, Centre for Climate Change Economics and Policy, Bruegel, 17 November 2016) <https://bruegel.org/wp-content/uploads/2016/12/Averchenkova-et-al-2016.pdf>.

Bahrij, John, and Lily Ko, 'An Overview of English Resources for Chinese Legal Research' (2013) 13 *Legal Information Management* 25.

Ball, Jeffrey, 'Grow Green China Inc. How China's Epic Push for Cleaner Energy Creates Economic Opportunity for the West' (Brookings, May 2019) <www.brookings.edu/research/grow-green-china-inc/>.

Barnett, Jon, 'Adapting to Climate Change in Pacific Island Countries: The Problem of Uncertainty' (2001) 29(6) *World Development* 977.

Bie, Tao, 'The Starting Point for Legislation of EPIL: Comment and Analysis for Amendment of the Civil Procedure Law and Advice on Amendment of the Environmental Protection Law' (环境公益诉讼立法的新起点，《民诉法》修改之评析与《环保法》修改之建议) (2013) 1 *Law Review* (法学评论) 101.

Bodansky, Daniel, 'The United Nations Framework Convention on Climate Change: A Commentary' (1993) 18(2) *Yale Journal of International Law* 451.

Bodenheimer, Edgar, *Jurisprudence: The Philosophy and Method of the Law*, 325 (Harvard University Press, 1981).

Bogojević, Sanja, *Emissions Trading Schemes: Markets, States and Law* (Hart Publishing, 2013).

Böhm, Steffen and Siddhartha Dabhi (eds), *Upsetting the Offset: The Political Economy of Carbon Markets* (MayFly Books, 2009).

Boute, Anatole and Hao Zhang, 'The Role of the Market and Traditional Regulation to Decarbonise China's Energy Supply' (2018) 30(2) *Journal of Environmental Law* 261.

BP, *Statistical Review of World Energy* (BP, June 2018) <www.bp.com/content/dam/bp/business-sites/en/global/corporate/pdfs/energy-economics/statistical-review/bp-stats-review-2018-full-report.pdf>.

Buckley, Chris, 'China Eyes Steel, Cement Sectors for Carbon Credit Trade-Report' *Reuters* (London, 4 August 2011) <www.reuters.com/article/2011/08/04/china-emissions-trading-idUSL3E7J407J20110804>.

Callon, Michel, 'Civilizing Markets: Carbon Trading between in vitro and in vivo Experiments' (2009) 34 *Accounting, Organizations and Society* 542.

Cames, Martin et al., 'How Additional is the Clean Development Mechanism? Analysis of the Application of Current Tools and Proposed Alternatives' (Öko-Institut, Study prepared for DG CLIMA, 2016) <https://ec.europa.eu/clima/sites/clima/files/ets/docs/clean_dev_mechanism_en.pdf>.

Cao, Mingde, 'The Legal Position and Strategy for China's Participation in Global Climate Governance' (中国参与国际气候治理的法律立场和策略 – 以气候正义为视角) (2016) 1 *China Legal Science* (中国法学) 29.

Cao, Wei and Li Yanfang, 'Framework of Laws and Policies on Renewable Energy and Relevant Systems in China under the Background of Climate Change' (2012) 13(4) *Vermont Journal of Environmental Law* 823.

Capannelli, Elisabetta and Omkar L. Shrestha, 'Environmental Challenges in the People's Republic of China and Scope for Bank Assistance' (Asian Development Bank, December 1993).

CAT, 'China: Overview' (Climate Action Tracker, 2 December 2019) <https://climateactiontracker.org/countries/china/>.

CAT, 'China: Pledges and Targets' (Climate Action Tracker, 2 December 2019) <https://climateactiontracker.org/countries/china/pledges-and-targets/>.

Center for Legal Assistance to Pollution Victims, 'Wang Canfa: The Practice of the EPIL Filed by Environmental NGOs and the Challenges It Faces' (王灿发：环保社会组织提起环境公益诉讼的实践与面临的挑战) (Center for Legal Assistance to Pollution Victims, 26 June 2017) <www.clapv.org/NewsContent.asp?id=1155&title=%D7%EE%D0%C2%B6%AF%CC%AC&titlecontent=NewsList&lei1 = 106>.

Chang, Jiwen, 'The Emission Control of CO_2 and the Amendment of Law on Air Pollution Prevention and Control' (二氧化碳的排放控制与《大气污染防治法》的修订) (2009) 3 *Law Science Magazine* (法学杂志) 74.

Chang, Yen-Chiang, Chuanliang Wang, Mehran Idris Khan, and Nannan Wang, 'Legal Practices and Challenges in Addressing Climate Change and Its Impact on the Oceans – a Chinese Perspective' (2020) 111 *Marine Policy* 103354.

Chayes, Abram, and Charlotte J. Kim, 'China and the United Nations Framework Convention on Climate Change', in Michael B. McElroy, Chris P. Nielsen, and Peter Lydon (eds), *Energizing China: Reconciling Environmental Protection and Economic Growth*, 503–40 (Harvard University Press, 1998).

Chen, Albert H. Y., 'China's Long March Towards Rule of Law or China's Turn against Law?' (2016) 4 *The Chinese Journal of Comparative Law* 1.

Chen, Gang, 'China's Solar PV Manufacturing and Subsidies from the Perspective of State Capitalism' (2015) 33(10) *Copenhagen Journal of Asian Studies* 90.

Chen, Haisong, 'Responding to Climate Change in Developing Countries: Legislation and Revelation' (发展中国家应对气候变化立法及其启示) (2013) 4 *Journal of Nanjing University of Technology (Social Science edition)* (南京工业大学学报(社会科学版)) 5.

Chen, Huizhen, 'Reduction Target and Total Emission Allowance: Experience and Implication of EU's Carbon Emission Scheme' (减排目标与总量设定:欧盟碳排放交易体系的经验及启示) (2013) 15(4) *Journal of Jiangsu University (Social Science Edition)* (江苏大学学报(社会科学版)) 14.

Chen, Huizhen, *Towards a Market-Based Climate Regime in China? A Legal Perspective on the Design and Implementation of Greenhouse Gas Emissions Trading* (Doctoral Dissertation, Maastricht University, 2015), 135.

Chen, Jianfu, *Chinese Law: Context and Transformation* (Revised and Expanded Edition) (Martinus Nijhoff, 2016).

Chen, Jinqiang, 'The Challenges and Promises of Greening China's Economy' (January 2017, Harvard Kennedy School, Belfer Center for Science and International Affairs) <www.belfercenter.org/publication/challenges-and-promises-greening-chinas-economy>.

Chen, Siyu, 'Integrating Adaptation to Climate Change in Project-level Environmental Impact Assessment' (论将适应气候变化的要求纳入建设项目环境影响评价制度) (2019) 33(4) *Journal of Chongqing University of Technology (Social Science)* (重庆理工大学学报(社会科学)) 27.

Chen, Tan, 'A Primary Discussion on Legitimation and Enactment of Policy' (浅论政策合法化与政策法律化行政与法) (2001) 1 *Public Administration & Law* (行政与法) 55.

Cheng, Xiaofeng, Hu Ke, and Jiang Xinyan, 'China', in Theodore L. Garrett (ed), *The Environment and Climate Change Law Review,* 42 (3rd edn, The Law Reviews, 2019).

Cheng, Xinhe, and Chen Huizhen, 'Two Dimensions of Carbon Trading Allowance: A Case Study of Guangdong Province' (论碳交易配额分配法律制度的两大维度 – 以广东为例) (2014) 2 *Journal of South China Normal University (Social Science Edition)* (华南师范大学学报(社会科学版)) 103.

Cheng, Yuyan, 'The Construction of a Legal Mechanism for Regional Cooperation of Climate Change Responses Among Local Governments' (地方政府应对气候变化区域合作的法治机制构建) (2016) 2 *Guangdong Social Science* (广东社会科学) 241.

Chinese Academy of Social Sciences Institute of Law and the Swiss Agency for Development and Cooperation, 'The Act of the People's Republic of China Addressing Climate Change (the Draft Proposal)' (2014) 2 *China Legal Science* 122.

Chiu, Dominic, 'The East Is Green: China's Global Leadership in Renewable Energy' (Center for Strategic and International Studies, 6 October 2017) <www.csis.org/east-green-chinas-global-leadership-renewable-energy>.

Chow, Daniel C. K., *The Legal System of the People's Republic of China in a Nutshell* (3rd edn, West Academic Publishing, 2015).

Christoffersen, Gaye, 'China's "Comprehensive" Energy Policy', in Thomas W. Robinson (ed), *The Foreign Relations of China's Environmental Policy,* 49 (The American Enterprise Institute (Conference Report), 1992).

Cimato, Federica, and Michael Mullan, *Adapting to Climate Change: Analysing the Role of Government* (Evidence and Analysis Series, Department for Environment Food and Rural Affairs, 2010).

Climate Action Tracker (CAT), 'China – Country Summary' (Climate Action Tracker, 2 December 2019) <https://climateactiontracker.org/countries/china/>.

Cliquet, An, Frank Maes, Willemien du Plessis, and Heather McLeod-Kilmurray, *Biodiversity and Climate Change* (Edward Elgar, 2013).

Cole, Daniel H., 'Origins of Emissions Trading in Theory and Early Practice', in Stefan E. Weishaar (ed), *Research Handbook on Emissions Trading,* 9–26 (Edward Elgar, 2016).

Conrad, Björn, 'China in Copenhagen: Reconciling the "Beijing Climate Revolution" and the "Copenhagen Climate Obstinacy"' (2012) 210 *The China Quarterly* 435.

Corfee-Morlot, Jan, 'Cities, Climate Change and Multilevel Governance' (OECD Environmental Working Papers, No. 14, 2009).

Cramton, Peter, David J. C. MacKay, Axel Ockenfels, and Steven Stoft (eds), *Global Carbon Pricing: The Path to Climate Cooperation* (MIT Press, 2017).

Dai, Rui, 'The SPC: 298 EPIL Cases Filed by NGOs Have Been Accepted Since 2015' (最高法：2015年以来受理社会组织提起环境公益诉讼案件298件) (The Cover, 30 July 2019) <www.thecover.cn/news/2410618>.

Davies, Paul A., and R. Andrew Westgate, 'The Chinese Government Announced Measures to Assist with ETS Implementation' (Latham & Watkins LLP, July 2018) <www.globalelr.com/2018/07/the-chinese-government-announced-measures-to-assist-with-ets-implementation/>.

De Cendra De Larragán, Javier, 'The Kyoto Protocol, with a Special Focus on the Flexible Mechanisms', in Daniel A. Farber and Marjan Peeters (eds), *Climate Change Law,* 227–238 (Elgar Encyclopedia of Environmental Law, Edward Elgar, 2016).

Deng, Haifeng, 'Legal Mechanisms for Implementing the Carbon Tax' (碳税实施的法律保障机制研究) (2014) 4 *Global Law Review* (环球法律评论) 104.

Deng, Haifeng, 'Systematic Outline of the Pollution Emission Right' (排污权制度论纲) (2007) 6 *Legal Science* (法律科学) 76.

Deng, Zhe, Dongya Li, Tao Pang, and Maosheng Duan, 'Effectiveness of Pilot Carbon Emissions Trading Systems in China' (2018) 18 *Climate Policy* 992.

Dent, Christopher, 'China's Renewable Energy Development: Policy, Industry and Business Perspectives' (2015) 21(1) *Asia Pacific Business Review* 26.

Department of Climate Change of the MEE, 'Progress of Constructing China's ETS and Considerations for Next Steps' (中国碳排放交易市场建设进展及下一步工作考虑) (2018) 46 *Environmental Protection* (环境保护) 14.

Diao, Fanchao, 'Lanzhou Intermediate People's Court Dismissed the Wind and Solar Curtailment Case FON v. State Grid Gansu Electric Power Co.: The Litigation Is not Eligible' (兰州中院驳回自然之友对国网甘肃公司弃风弃光诉讼：条件不符) (The Paper, 20 August 2018) <www.thepaper.cn/newsDetail_forward_2364374>.

Diao, Fanchao, 'NGO Claims RMB 0.3 Billion for Wind and Solar Curtailment; Ningxia Grid: We Are Not Able to Purchase the Electricity in Full Amount' ("弃风弃光"被环保组织索赔3亿，宁夏电网：做不到全额收购) (The Paper, 11 April 2018) <www.thepaper.cn/newsDetail_forward_2069105>.

Ding, Ding, and Pan Fangfang, 'Legal Attributes of Carbon Emission Right' (论碳排放权的法律属性) (2012) 9 *Legal Science Magazine* (法学杂志) 105.

DiPeso, Jim, 'Carbon Offsets: Is the Environment Getting What You Pay For?' (2007) 17 *Environmental Quality Management* 89.

Duan, Maosheng, Tao Pang, and Xiliang Zhang, 'Review of Carbon Emissions Trading Pilots in China' (2014) 25(3) *Energy and Environment* 527.

Duan, Maosheng, and Li Zhou, 'Key Issues in Designing China's National Carbon Emissions Trading System' (2017) 6(2) *Economics of Energy and Environmental Policy* 55.

Economy, Elizabeth, 'Chinese Policy-making and Global Climate Change: Two-front Diplomacy and the International Community', in Miranda A. Schreurs and Elizabeth Economy (eds), *The Internationalization of Environmental Protection*, 19–41 (Cambridge University Press, 1997).

Eisen, Joel, 'China's Renewable Energy Law: A Platform for Green Leadership?' (2010) 35(1) *William and Mary Environmental Law and Policy Review* 1.

Energy Charter Secretariat, 'China Energy Efficiency Report: Protocol on Energy Efficiency and Environmental Aspects' (Energy Charter Secretariat, 2018).

Energy Research Institute, China Law Society, *China Energy Law Research Report (2009)* (中国能源法研究报告(2009)) (Law Press China, 2010).

Energy Research Institute of Academy of Macroeconomic Research and China National Renewable Energy Centre, 'China Renewable Energy Outlook 2017' (10 November 2017) <http://boostre.cnrec.org.cn/wp-content/uploads/2017/10/CREO-2017-EN-20171113-1.pdf>.

Etzioni, Amitai, 'Is Transparency the Best Disinfectant?' (2010) 18(4) *The Journal of Political Philosophy* 389.

Falconer, Angela et al., 'Tracking Emissions and Mitigation Actions: Evaluation of MRV Systems in China, Germany, Italy, and the United States' (San Francisco, Climate Policy Initiative, 2012).

Fan, Jinglin et al., 'The Development of China's Renewable Energy Policy and Implications for Africa' (2018) 394(4) *IOP Conference Series Materials: Science and Engineering* 2.

Fang, Dongsheng, 'Reconsideration of Relation between Law and Policy—Empiricism as the Angle of View' (法律与政策关系的再认识 – 以实证为视角) (2008) 29 *Journal of Jining University* (济宁学院学报) 87.

Fang, Mandy Meng, 'A Crisis or An Opportunity? The Trade War between the US and China in the Solar PV Sector' (2020) *Journal of World Trade* forthcoming.

Fang, Shirong, and Sun Caihua, 'Reconsideration of the Relationship between Policies and Law – from the Viewpoint of Concentrating and Reflecting the Will of the People' (关于政策与法律关系的再思考 – 从集中和反映人民意志的视角) (2007) 5 *Hubei Social Sciences* (湖北社会科学) 17.

Felicity Deane, Evan Hamman, and Yilin Pei, 'Principles of Transparency in Emissions Trading Schemes: The Chinese Experience' (2017) 6(1) *Transnational Environmental Law* 87.

Feng, Chuan, Leyton P. Nelson, and Thomas W. Simon, *China's Changing Legal System: Lawyers and Judges on Civil and Criminal Law* (Palgrave Macmillan, 2016).

Feng, Hao, and Tom Baxter, 'China's Coal Consumption on the Rise' (China Dialogue, Beijing, 1 March 2019) <www.chinadialogue.net/article/show/single/en/11107-China-s-coal-consumption-on-the-rise>.

Finamore, Barbara, 'China Renews Its Commitment to Renewable Energy' (1 February 2010, Natural Resources Defense Council) <www.nrdc.org/experts/barbara-finamore/china-renews-its-commitment-renewable-energy>.

Fischman, Robert, and Jillian R. Rountree, 'Adaptive Management', in Michael B. Gerrard and Katrina Fischer Kuh (eds), *The Law of Adaptation to Climate Change: US and International Aspects,* 5–29 (American Bar Association, 2012).

Fransen, Taryn et al., 'Enhancing NDCs: A Guide to Strengthening National Climate Plans by 2020' (Washington, DC, World Resources Institute, 2019).

Friedrich, Johannes, Ge Mengpin, and Andrew Pickens, 'This Interactive Chart Explains World's Top 10 Emitters, and How They've Changed' (World Resources

Institute, 11 April 2017) <www.wri.org/blog/2017/04/interactive-chart-explains-worlds-top-10-emitters-and-how-theyve-changed>.

Friends of Nature (FON), 'Annual Report 2018' (自然之友2018年报) (FON, 19 April 2019) <www.fon.org.cn/index.php?option=com_k2&view=item&id=13554:2018&Itemid=119>.

Friends of Nature (FON), 'Gansu Wind Curtailment Case: We Start Again' (甘肃弃风案，我们"重头再来") (FON, 28 January 2019) <www.fon.org.cn/index.php?option=com_k2&view=item&id=13433:2019-01-28-03-10-35&Itemid=176>.

Friends of Nature (FON), 'Guangdong Nanling: the Ecological Remediation Is Ongoing. Thanks for the Steady Efforts' (广东南岭：生态修复进行时，感谢所有一步一个脚印的努力) (FON, 25 June 2018) <www.fon.org.cn/index.php?option=com_k2&view=item&id=13109:2018-06-25-03-43-27&Itemid=176>.

Gallagher, Kelly Sims et al., 'Assessing the Policy Gaps for Achieving China's Climate Targets in the Paris Agreement' (2019) 10 *Nature Communications* 1.

Gao, Xiaosheng, 'The Paris Agreement and Global Climate Governance: China's Role and Contribution' (2016) 2 *China Quarterly of International Strategic Studies* 365.

Gao, Zhiwen, 'Analysis of CCER Restrictive Policies in China's Carbon Emission Trading Pilots' (中国各碳排放权交易试点CCER限制性政策分析) (Climate Bridge, 2014) <www.climatebridge.com/news/DownLoad.aspx?ID=/uploads/20150918/20150918160948367.pdf>.

Garnaut, Ross, *The Garnaut Climate Change Review: Final Report* (Cambridge University Press, 2008).

Gilbertson, Tamra, and Oscar Reyes, 'Carbon Trading: How It Works and Why It Fails' (Dag Hammarskjöld Foundation, 2009) <www.daghammarskjold.se/wp-content/uploads/2014/08/cc7_web_low.pdf>.

Gillenwater, Michael, and Stephen Seres, 'The CDM: A Review of the First International Offset Program' (2011) 3–4 *Greenhouse Gas Measurement & Management* 179.

Global Environment Facility, 'Annual Performance Report 2019' <www.thegef.org/sites/default/files/council-meeting-documents/EN_GEF.ME_C56_Inf.01_Annual_Performance_Report_May_2019_0.pdf>.

Gong, Wei, 'Cooperative Control of Air Pollutants and GHGs: Challenges and Responses from the Perspective of Enforcement' (大气污染物与温室气体协同控制面临的挑战与应对 – 以法律实施为视角) (2017) 1 *Journal of Southwestern University for Nationalities (Humanities and Social Science Edition)* (西南民族大学学报) (人文社会科学版) 108.

Gong, Wei, 'International Law on Climate Change and the Legislative Model of China's Climate Change Response' (气候变化国际法与我国气候变化立法模式) (2013) 37(3) *Journal of Xiangtan University* (湘潭大学学报) 42.

Goron, Coraline, and Cyril Cassisa, 'Regulatory Institutions and Market-Based Climate Policy in China' (2017) 17 *Global Environmental Politics* 99.

Green, Fergus, and Nicholas Stern, 'China's Changing Economy: Implications for its Carbon Dioxide Emissions' (2017) 17(4) *Climate Policy* 423.

Guo, Dongmei, 'Analysis of Legal Strategy against Climate Change: From Outer System to Inner Rules' (气候变化法律应对实证分析 – 从国际公约到国内法的转化) (2010) 12 *Journal of Southwest University of Political Science & Law* (西南政法大学学报) 45.

Guo, Dongmei, *Studies on the Legal System in Relation to Climate Change* (应对气候变化法律制度研究) (Law Press, 2010).

Gupta, Aarti, and Michael Mason, 'Transparency', in Karin Bäckstrand and Eva Lövbrand (eds), *Research Handbook on Climate Governance,* 446–457 (Cheltenham, UK: Elgar, 2015).

Handmer, J. W., S. Dovers, and T. E. Downing, 'Social Vulnerability to Climate Change and Variability' (1999) 4(3–4) *Mitigation and Adaptation Strategies for Global Change* 267.

Hao, Yufan, 'Environmental Protection in Chinese Foreign Policy' (1992) 11(3) *Journal of Northeast Asian Studies* 25.

Hatch, Michael T., 'Chinese Politics, Energy Policy, and the International Climate Change Negotiations', in Paul G. Harris (ed), *Global Warming and East Asia: The Domestic and International Politics of Climate Change,* 43–66 (Routledge, 2003).

He, Xiangbai, 'China's Legal Framework on Climate Change Adaptation: Ambitious or Conservative' (突破还是保守:评我国适应气候变化的法律框架) (2015) 2 *Nanjing University Law Review* (南京大学法律评论) 230.

He, Xiangbai, 'Chinese Local Governments to be Held Responsible for Their Local Environment: New Law, Old Problems' (2019) 17(3) *China: An International Journal* 28.

He, Xiangbai, 'Legal and Policy Pathways of Climate Change Adaptation: Comparative Analysis of the Adaptation Practices in the United States, Australia and China' (2018) 7(2) *Transnational Environmental Law* 347.

He, Xiangbai, *Legal Methods of Mainstreaming Climate Change Adaptation in Chinese Water Management* (Springer, 2016).

He, Ying, 'Legislative Regulation of Carbon Emission Trading' (我国碳排放权交易立法规制思考) (2018) 2 *Journal of Southern China Normal University (Social Science Edition)* (华南师范大学学报(社会科学版) 157.

High-Level Commission on Carbon Prices, *Report of the High-Level Commission on Carbon Prices* (Washington, DC, World Bank, 2017).

Hilson, Chris, 'Climate Change Litigation in the UK: An Explanatory Approach (or Bringing Grievance Back In)', in F. Fracchia and M. Occhiena (eds), *Climate Change: La Riposta del Diritto,* 421–436 (Editoriale Scientifica, 2010).

Hilton, Isabel, and Oliver Kerr, 'The Paris Agreement: China's "New Normal" Role in International Climate Negotiations' (2017) 17(1) *Climate Policy* 48.

Ho, Mun, Zhongmin Wang, and Zichao Yu, 'China's Power Generation Dispatch' (Resources for the Future 2017) <www.rff.org/files/document/file/RFF-Rpt-ChinaElectricity.pdf>.

Hogan, Pat et al., 'Tracking Emissions and Mitigation Actions: Current Practice in China, Germany, Italy, and the United States' (San Francisco, Climate Policy Initiative, 2012), 10.

Hoogzaad, Jelmer, and Charlotte Streck, 'A Mechanism With a Bright Future: Joint Implementation', in David Freestone and Charlotte Streck (eds), *Legal Aspects of Carbon Trading: Kyoto, Copenhagen, and Beyond,* 176–194 (Oxford University Press, 2009).

Houghton, J. T. et al. (eds), *Climate Change 2001: The Scientific Basis* (Cambridge University Press, 2001).

Hove, Anders, and Daniel Wetzel, 'China is Planning Provincial Quotas for Clean Energy' (China Dialogue, 2018) <www.chinadialogue.net/blog/10574-China-is-planning-provincial-quotas-for-clean-energy-/en>.

Hsu, Angel, Alex de Sherbinin, and Han Shi, 'Seeking Truth from Facts: The Challenge of Environmental Indicator Development in China' (2012) (39)3 *Environmental Development* 40.

Hu, Bin, and Gao Fang, 'Carbon Trading or Carbon Tax?' (碳交易还是碳税) (2015) 4 *World Environment* (世界环境) 50.

Hu, Yuan, and Zheng Shaohua, 'From Authoritarian Management to Social Governance: Reflections on the Amendment of Law on Air Pollution Prevention and Control' (从威权管制到社会治理 - 关于修订《大气污染防治法》的几点思考) (2010) 6 *Modern Law Science* (现代法学) 150.

Huang, Jinrong, 'Legal Definition and Legal Nature of Normative Documents' ("规范性文件" 的法律界定及其效力) (2014) 7 *Legal Science* (法学) 12.

Huang, Ping et al., 'How China Became a Leader in Solar PV: An Innovation System Analysis' (2016) 64 *Renewable and Sustainable Energy Reviews* 777.

International Energy Agency, 'Global Energy and CO_2 Status Report 2018: The Latest Trends in Energy and Emissions in 2018' <https://webstore.iea.org/global-energy-co2-status-report-2018>.

IPCC, *Climate Change 2014: Impacts, Adaptation, and Vulnerability: Contribution of Working Group II to the Fifth Assessment Report of the Intergovernmental Panel on Climate Change* (IPCC, 2014).

IPCC, *Good Practice Guidance and Uncertainty Management in National Greenhouse Gas Inventories* (IPCC, 2000).

IPCC, *Good Practice Guidance for Land Use, Land-Use Change and Forestry* (IPCC, 2003).

IPCC, *Guidelines for National Greenhouse Gas Inventories* (IPCC, 2006).

IPCC, *Revised 1996 IPCC Guidelines for National Greenhouse Gas Inventories* (IPCC, 1996).

IRENA, '30 Years of Policies for Wind Energy Lessons from 12 Wind Energy Markets' (IRENA, 2013) <www.irena.org/documentdownloads/publications/gwec_china.pdf>.

IRENA, 'Renewable Energy Benefits: Measuring the Economics' (January 2016) <www.irena.org/publications/2016/Jan/Renewable-Energy-Benefits-Measuring-the-Economics>.

IRENA, 'Renewable Energy Prospects: China' (November 2014) <www.irena.org/publications/2014/Nov/Renewable-Energy-Prospects-China>.

IRENA, 'Renewable Energy Prospects: China, RE Map 2030 Analysis' (November 2014) <www.irena.org/-/media/Files/IRENA/Agency/Publication/2014/IRENA_REmap_China_report_2014.pdf>.

Ji, Yaping, 'Studies on Public Participation in Administrative Decision-Making' (行政决策程序中的公众参与研究) (2012) 8 *Zhejiang Academic Journal* (浙江学刊) 166.

Jia, Mark, 'Chinese Common Law? Guiding Cases and Judicial Reform' (2016) 129 *Harvard Law Review* 2213.

Jiang, Jing Jing, Bin Ye, and Xiao Ming Ma, 'The Construction of Shenzhen's Carbon Emission Trading Scheme' (2014) 75 *Energy Policy* 17.

Jiang, Xiaoyi, *Legal Issues for Implementing the Clean Development Mechanism in China* (Springer, 2013).

Johnston, Alastair Iain, 'China and International Environmental Institutions: A Decision Rule Analysis', in Michael B. McElroy, Chris P. Nielsen, and Peter Lydon (eds), *Energizing China: Reconciling Environmental Protection and Economic Growth,* 555–600 (Harvard University Press, 1998).

Jones, Judith, 'Regulatory Design for Scientific Uncertainty: Acknowledging the Diversity of Approaches in Environmental Regulation and Public Administration' (2007) 19(3) *Journal of Environmental Law* 347.

Jordan, Andrew et al., 'Emergence of Polycentric Climate Governance and Its Future Prospects' (2015) 5 *Nature Climate Change* 977.

Jotzo, Frank, and Andreas Löschel, 'Emissions Trading in China: Emerging Experiences and International Lessons' (2014) 75 *Energy Policy* 3.

Kahrl, Fredrich, James H. Williams, and Junfeng Hu, 'The Political Economy of Electricity Dispatch Reform in China' (2013) 53 *Energy Policy* 361.

Kahrl, Fredrich, Jim Williams, Ding Jianhua, and Junfeng Hu, 'Challenges to China's Transition to a Low Carbon Electricity System' (2011) 39 *Energy Policy* 4032.

Knight, Zoe, Nick Robins, and Wai-Shin Chan, 'Scoping Climate Change Risks: Which Countries are Most Vulnerable?' (HSBC Global Research Report, HSBC, 2011).

Koch-Weser, Iacob, 'China's Wind and Solar Sectors: Trends in Deployment, Manufacturing, and Energy Policy' (US – China Economic and Security Review Commission Staff Research Report, 9 March 2015).

Koehn, Peter H., *China Confronts Climate Change: A Bottom-up Perspective* (Routledge, 2015).

Korsbakken, Jan Ivar, Glen P. Peters, and Robbie M. Andrew, 'Uncertainties around Reductions in China's Coal Use and CO_2 Emissions' (2016) 6 *Nature Climate Change* 687.

Kossof, Paul, *Chinese Legal Research* (Carolina Academic Press, 2014).

Kostka, Genia, 'Command without Control: The Case of China's Environmental Target System' (2016) 10 *Regulation & Governance* 58.

Kurukulasuriya, Lal, and Nicholas A. Robinson, *Training Manual on International Environmental Law* (UNEP/Earthprint, 2006).

Lang, Huanlin, *Public Participation in Environmental Decision-Making in China: Towards an Ecosystem Approach* (Doctoral Dissertation, University of Groningen, 2014).

Le Quéré, Corinne et al., 'Drivers of Declining CO_2 Emissions in 18 Developed Economies' (2019) 9 *Nature Climate Change* 213.

Lederer, Markus, 'Market Making via Regulation: The Role of the State in Carbon Markets' (2012) 6 *Regulation & Governance* 526.

Lee, Godden, and Jacqueline Peel, *Environmental Law: Science, Policy and Regulatory Dimensions* (Oxford University Press, 2010).

Leggett, Jane A., 'China's Greenhouse Gas Emissions and Mitigation Policies' (CRS Report for Congress No. R41919, Congressional Research Service, 2011).

Legrand, Pierre, 'The Impossibility of "Legal Transplants"' (1997) 4 *Maastricht Journal of European & Comparative Law* 111.

Leitch, Anne, Ben Harman, and Marcus B. Lane, 'From Blueprint to Footprint: Climate Change and the Challenge for Planning', in Tim Bonyhady, Andrew Macintosh, and Jan McDonald (eds), *Adaptation to Climate Change: Law and Policy,* 63–79 (Federation Press, 2010).

Lesnikowski, Alexandra et al., 'What Does the Paris Agreement Mean for Adaptation?' (2017) 17(7) *Climate Policy* 825–831.

Leung, Guy, 'China's Energy Security: Perception and Reality' (2011) 39(3) *Energy Policy* 1330.

Lewis, Joanna, 'Building A National Wind Turbine Industry: Experiences from China, India and South Korea' (2011) 5(3/4) *International Journal of Technology and Globalization* 281.

Li, Anthony H. F., 'Hopes of Limiting Global Warming? China and the Paris Agreement on Climate Change' (2016) 1 *China Perspectives* 49.

Li, Junfeng et al., *China Wind Power Report* (China Environmental Science Press, 2007).

Li, Lin, *Building the Rule of Law in China* (Elsevier Asian Studies Series, Chandos Publishing, 2017).

Li, Shuo et al., 'Main Focus of Adaptation Under the Implementation Rules of the Paris Agreement and China's Future Measures' (《巴黎协定》实施细则适应议题焦点解析及后续中国应对措施) (2019) 15(4) *Climate Change Research* (气候变化研究进展) 436.

Li, Yanfang, 'Comparison on National Legislation on Climate Change and Its Implications for China' (各国应对气候变化立法比较及其对中国的启示) (2010) 4 *Journal of Renmin University of China* (中国人民大学学报) 58.

Li, Yanfang, 'On the Establishment of a Legal System for China's Reaction to Climate Change' (论中国应对气候变化法律体系的建立) (2010) *Journal of China University of Political Science and Law* (中国政法大学学报) 84.

Li, Yanfang, and Tian Shiyu, 'Climate Change Regulatory Risk under the Context of Uncertainty and Complexity' (不确定性与复杂性背景下气候变化风险规制立法) (2018) 58(2) *Journal of Jilin University Social Science* (吉林大学社会科学学报) 42.

Li, Yanfang, and Zhang Zhongli, 'Legal Status of CO_2 and the Legislative Approach of Its Emission Regulation' (二氧化碳的法律定位及其排放规制立法路径选择) (2015) 2 *Social Science Study* (社会科学研究) 108.

Li, Yanfang, Zhang Zhongli, and Li Cheng, 'Toward Specialized Legislation to Address Climate Change in China' (我国应对气候变化立法的若干思考) (2016) 33 *Journal of Shanghai University (Social Sciences)* (上海大学学报: 社会科学版) 8.

Li, Yanfang et al., 'Thoughts on China's Climate Change Legislation' (我国应对气候变化立法的若干思考) (2016) 1 *Journal of Shanghai University (Social Science Edition)* (上海大学学报(社会科学版)) 1.

Li, Yumei, 'Study of China's Climate Change Legislation' (我国气候变化立法刍议) (2015) 33(1) *Tribune of Political Science and Law* (政法论坛) 182.

Li, Zhiping, 'Supervision Mechanism of Carbon Trading Market' (碳交易市场的监管机制研究) (2012) 14(1) *Journal of Jiangsu University (Social Science Edition)* (江苏大学学报(社会科学版)) 56.

Lin, Yan, 'Constitutional Evolution through Legislation: The Quiet Transformation of China's Constitution' (2015) 13 *International Journal of Constitutional Law* 61.

Liu, Chang, 'Study on the Sustainability of the Effect of Energy-Saving Policy' (节能政策效果的持续性探究) (2014) 5 *Economic Review* (经济纵横) 25.

Liu, Chen, and Yao Song, 'Progress of China's ETS and Some Recommendations' (中国碳排放权交易市场建设现状与建议) (2019) 4 *International Petroleum Economics* (国际石油经济) 47.

Liu, Jing, 'The Legality of Greenhouse Gas Emission Reduction: A Coordinated Control of Greenhouse Gases and Air Pollutants – a Comment on Article 2, Paragraph 2, of the Law on Prevention and Control of Atmospheric Pollution' (温室气体减排的法律路径：温室气体和大气污染物协同控制 – 评《大气污染防治法》第2条第2款) (2019) 47 *Journal of Xinjiang University (Philosophy, Humanities & Social Science)* (新疆大学学报(哲学·人文社会科学版)) 52.

Liu, Liqun et al., 'Solar Energy Development in China – A Review' (2010) 14(1) *Renewable and Sustainable Energy Reviews* 301.

Liu, Mingming, 'The Regulation of GHG Emissions: Content and Boundary' (论温室气体排放管制的内涵及边界) (2012) 18 *Journal of Central South University (Social Science Edition)* (中南大学学报(社会科学版)) 105.

Liu, Shen, and Zhang Yingtao, 'On the Principle of Subsidiarity and the Reform of China's Administrative System' (辅助性原则与中国行政体制改革) (2006) 4 *Administrative Law Review* (行政法学研究) 9.

Liu, Shiyu et al., 'Curtailment of Renewable Energy in Northwest China and Market-based Solutions' (2018) 123 *Energy Policy* 494.

Liu, Siqi, 'The Status Quo, Problem Analysis and Countermeasures Study of China's Carbon Emission Trading' (中国碳排放交易试点的现状、问题分析及对策研究) (2015) *China's Environmental Rule of Law* (中国环境法治) 27.

Liu, Songshan, 'The Criterion the Court Relies on for Trial' (人民法院的审判依据) (2006) 24 *Tribune of Political Science and Law: Journal of China University of Political Science and Law* (政法论坛: 中国政法大学学报) 22.

Liu, Zhe, 'The Basic Problems in China's Legislation on Climate Change' (我国应对气候变化立法的基本问题探究) (2016) 18 *Journal of Jiangsu University (Social Science Edition)* (江苏大学学报(社会科学版)) 16.

Liu, Zuoxiang, 'Norm Systems in Contemporary China: Theory and Institutional Structure' (当代中国的规范体系:理论与制度结构) (2019) 7 *Social Sciences in China* (中国社会科学) 96.

Lo, Alex Y., and Ren Cong, 'After CDM: Domestic Carbon Offsetting in China' (2017) 141 *Journal of Cleaner Production* 1391.

Lo, Kevin, 'A Critical Review of China's Rapidly Developing Renewable Energy and Energy Efficiency Polices' (2014) 29 *Renewable and Sustainable Energy Reviews* 508.

Locatelli, Catherine, 'China's Energy Policy: Energy and Economic Development' (1989) 1(2) *Energy Studies Review* 144.

Lovell, Heather C., 'Governing the Carbon Offset Market' (2010) 1 *Wiley Interdisciplinary Reviews: Climate Change* 353.

Lü, Jiang, and Liang Xiaofei, 'Core National Interests and Climate Change Legislation: Between Principles and Norms' (国家核心利益与气候变化立法:在原则与规范之间) (2014) 35(4) *Journal of Jishou University (Social Science Edition)* (吉首大学学报(社会科学版) 34.

Lü, Zhongmei, and Wang Guofei, 'The Development of China's Carbon Trading Market: Judicial Issues and Responses' (中国碳排放市场建设:司法问题及对策) (2016) 5 *Gansu Social Science* (甘肃社会科学) 161.

Luo, Li, 'Climate Change Legislation in Japan' (日本应对气候变化立法研究) (2010) 5 *Law Review* (法学论坛) 107.

Ma, Rongzhen, 'A "Green Peacock Defence Battle" Initiated by Many ENGOs' (一场多家环保NGO掀起的"绿孔雀保卫战") *Southern Weekly* (30 September 2018) <www.infzm.com/content/139882>.

Ma, Tianjie, 'China's Environment in 2015: A Year in Review' (China Dialogue, 23 December 2015) <www.chinadialogue.net/article/show/single/en/8497-China-s-environment-in-2015-a-year-in-review>.

MacIntosh, Andrew, 'A Theoretical Framework for Adaptation Policy', in Tim Bonyhady, Andrew Macintosh, and Jan McDonald (eds), *Adaptation to Climate Change: Law and Policy*, 38–62 (Federation Press, 2010).

Maraseni, Tek Narayan, and Gao Xinquan, 'An Analysis of Chinese Perceptions on Unilateral Clean Development Mechanism (uCDM) Projects' (2011) 14(3) *Environmental Science & Policy* 339.

Markell, David, and J. B. Ruhl, 'An Empirical Survey of Climate Change Litigation in the United States' (2010) 40 *Environmental Law Reporter* 10644.

Marks, Danny, 'China's Climate Change Policy Process: Improved but still Weak and Fragmented' (2010) 19(67) *Journal of Contemporary China* 971.

Mattews, John, and A. J. Wickel, 'Embracing Uncertainty in Freshwater Climate Change Adaptation: A Natural History Approach' (2009) 1(3) *Climate and Development* 269.

Mayer, Benoit, Mikko Rajavuori, and Fang Meng, 'The Contribution of State-Owned Enterprises to Climate Change Mitigation in China' (2017) 7(2–3) *Climate Law* 97.

McDonald, Jan, 'A Short History of Climate Adaptation Law in Australia' (2014) 4 *Climate Law* 150.

McGee, Henry W., 'Roscoe Pound's Legacy: Engineering Liberty and Order' (1970) 16 *Howard Law Journal* 19.

Meng, Wei, 'Integrating GHG Emissions in the Law on Air Pollution Prevention and Control' (孟伟：将温室气体排放问题纳入《大气污染防治法》) (Ministry of Natural Resources, 11 March 2010) <www.mnr.gov.cn/zt/hu/lh/2010qglh/lhhzy/gzhj/201003/t20100313_2022743.html>.

Miao, Yan, 'Jurisprudential Reflection and Structure Optimization of the Judicial Interpretation System' (司法解释制度之法理反思与结构优化) (2019) 25 *Law and Social Development* (法制与社会发展) 96.

Moore, Scott, 'Strategic Imperative? Reading China's Climate Policy in Terms of Core Interests' (2011) 23(2) *Global Change, Peace and Security* 147.

Morgan, Jennifer, and Deborah Seligsohn, 'What Cancun Means for China and the U.S.' (World Resources Institute, 16 December 2010).

Mulvaney, Kieran, 'Climate Change Report Card: These Countries Are Reaching Targets' (National Geographic, 19 September 2019) <www.nationalgeographic.com/environment/2019/09/climate-change-report-card-co2-emissions>.

Neelam Singh, Jared Finnegan, and Kelly Levin, 'MRV 101: Understanding Measurement, Reporting, and Verification of Climate Change Mitigation' (World Resources Institute, 2016), 4–5.

Neuweg, Isabella, and Alina Averchenkova, 'Climate Change Legislation and Policy in China, the European Union and the United States', in A. Averchenkova, S. Fankhauser, and M. Nachmany (eds), *Trends in Climate Change Legislation,* 37–59 (Edward Elgar, 2017).

Ni, Chunchun, 'China's Wind Power Generation Policy and Market Developments' (Institute of Energy Economics, Japan, 2008).

Nielsen, Chris P., and Michael B. McElroy, 'Introduction and Overview', in Michael B. McElroy, Chris P. Nielsen, and Peter Lydon (eds), *Energizing China: Reconciling Environmental Protection and Economic Growth,* 1–63 (Harvard University Press, 1998).

O'Brien, Bernie, and Mark Sculpher, 'Building Uncertainty into Cost-Effectiveness Rankings, Portfolio Risk-return Tradeoffs and Implications for Decision Rules' (2000) 38 *Medical Care* 460.

O'Brien, Roderick, 'Comparative Law and China's Socialist Legal System' (2014) 2014 *Journal of South African Law* 130.

Ohshita, Stephanie, Lynn Price, and Tian Zhiyu, 'The Available Target Allocation Method in Chinese Provinces: Energy Intensity During the Twelfth Five-Year Plan Period' (中国省区可用的目标分配方法：第十二个五年规划的能源强度) (Ernest Orlando Lawrence Berkeley National Laboratory, March 2011) <https://china.lbl.gov/sites/all/files/12th_fyp_target_allocation_report_cn_final_032111_0.pdf>.

Oksenberg, Michel, and Elizabeth Economy, 'Introduction: China Joins the World', in Elizabeth Economy and Michel Oksenberg (eds), *China Joins the World: Progress and Prospects,* 1–41 (Council on Foreign Relations Press, 1999).

Organization for Economic Cooperation and Development, *Integrating Climate Change Adaptation into Development Co-operation: Policy Guidance* (OECD, 2009).

Palmer, Doug, 'US Challenges China Wind Power Aid at WTO' (23 December 2010, Reuters) <www.reuters.com/article/us-usa-china-windpower/u-s-challenges-china-wind-power-aid-at-wto-idUSTRE6BL3EU20101222>.

Pan, Kexi et al., 'Estimation of Coal-Related CO_2 Emissions: The Case of China' (2013) 24(7–8) *Energy and Environment* 1309.

Pan, Xiaobin, 'Construction of Climate Change Law System in China' (中国应对气候变化法律体系的构建) (2016) 6 *Nankai Journal (Philosophy, Literature and Social Science Edition)* (南开学报:哲学社会科学版) 78.

Pan, Xiaobin, 'Litigation of China's Carbon Emission Trading' (我国碳排放交易法律诉讼问题) (2016) 4 *Tianjin Legal Science* (天津法学) 32.

Pan, Xiaobin, 'Research on China's First Legislation to Address Climate Change' (中国地方应对气候变化先行立法研究) (2017) 38 *Law Science Magazine* (法学杂志) 136.

Pan, Xiaobin, 'Study of Local Level Legislation on Climate Change in China' (中国地方应对气候变化先行立法研究) (2013) 3 *Law Science Magazine* (法学杂志) 132.

Pan, Xiaobin, 'The Construction of China's Legal System on Climate Change' (中国应对气候变化法律体系的构建) (2016) 6 *Nankai Journal (Philosophy and Social Science Edition)* (《南开学报》(哲学社会科学版)) 78.

Peel, Jacqueline, and Hari M. Osofsky, 'Climate Change Litigation's Regulatory Pathways: A Comparative Analysis of the United States and Australia' (2013) 35 *Law & Policy* 150.

Peeters, Marjan, Huizhen Chen, and Zhiping Li, 'Contrasting Emission Trading in the EU and China: An Exploration of the Role of the Courts' (2016) 6 *Climate Law* 197.

Peng, Benli, and He Yan, 'Climate Change Legislation in China: Features, Limitations and Prospects' (中国应对气候变化地方立法 – 特点、局限与展望) (2017) 6 *Journal of Yulin Normal College (Philosophy and Social Science Edition)* (玉林师范学院学报(哲学社会科学) 65.

Peng, Benli, and He Yan, 'Legislation at the Local Level to Address Climate Change in China: Features, Limitations and Prospects' (中国应对气候变化地方立法 – 特点、局限与展望) (2017) 38 *Journal of Yulin Normal University (Philosophy & Social Science)* (玉林师范学院学报(哲学社会科学)) 67.

Peng, Feng, and Yan Lidong, 'Comparative Study of Measurement, Reporting and Verification of Local Carbon Trading Pilots' (地方碳交易试点之"可测量、可报告、可核实制度"比较研究) (2015) 15(4) *Journal of China University of Geosciences* (中国地质大学学报) 26.

Peng, Feng, and Shao Shiyang, 'Carbon Emission Trading in EU: New Developments and Implications for China' (欧盟碳排放交易制度:最新动向及对中国之镜鉴) (2012) 12(5) *Journal of China University of Geosciences* (中国地质大学学报) 41.

Peng, Pu, 'Transforming the Old Mindset and Adjusting the Structure to Reduce Carbon Intensity' (转观念调结构 降低碳强度) (China Energy News, 17 February 2014).

Peng, Sizhen et al., 'Current Status, Problems and Recommendations on Climate Change Adaptation Policies in China' (中国适应气候变化政策现状、问题和建议) (2015) 25(9) *China Population, Resources and Environment* (中国人口·资源与环境) 4.

Perlack, Robert D., Milton Russell, and Zhongmin Shen, 'Reducing Greenhouse Gas Emissions in China: Institutional, Legal and Cultural Constraints and Opportunities' (1993) 3(1) *Global Environmental Change* 78.

Pielke, Roger et al., 'Climate Change 2007: Lifting the Taboo on Adaptation' (2007) 445 *Nature* 597.

Pound, Roscoe, *Social Control through Law* (Transaction Publishers, 1997).

Putnam, Robert D., 'Diplomacy and Domestic Politics: The Logic of Two-Level Games' (1998) 42(3) *International Organization* 427.

Qi, Shaozhou, Banban Wang, and Jihong Jiang, 'Policy Design of the Hubei ETS Pilot in China' (2014) 75 *Energy Policy* 31.

Qi, Ye, and Xiliang Zhang, *Annual Review of Low-Carbon Development in China 2018* (中国低碳发展报告2018) (Social Science Academic Publishing, 2018).

Qin, Dahe, 'Developing Legal System to Respond to Climate Change' (完善法律制度，应对气候变化) *Guang Ming Daily* (20 April 2009).

Qu, Geping, 'China's Dual-Thrust Energy Strategy: Economic Development and Environmental Protection' (1992) 20(6) *Energy Policy* 500.

Rajamani, Lavanya, 'The United Nations Framework Convention on Climate Change: A Framework Approach to Climate Change', in Daniel A. Farber and Marjan Peeters (eds), *Climate Change Law,* 205–216 (Elgar Encyclopedia of Environmental Law, Edward Elgar, 2016).

Refsgaard, J. C., 'The Role of Uncertainty in Climate Change Adaptation Strategies: A Danish Water Management Example' (2013) 18 *Mitigation and Adaptation Strategies for Global Change* 337.

Romain, Weikmans, Harro van Asselt, and J. Timmons Roberts, 'Transparency Requirements under the Paris Agreement and Their (Un)Likely Impact on Strengthening the Ambition of Nationally Determined Contributions (NDCs)' (2020) 20(4) *Climate Policy* 511.

Ross, Lester, 'The Politics of Environmental Policy in the People's Republic of China' (1992) 20(4) *Policy Studies Journal* 628.

Rotmans, Jan, 'Global Change and Sustainable Development: A Modelling Perspective for the Next Decade' (GLOBO Report Series No 4, Global Dynamics and Sustainable Development Programme, 1994).

Sautter, John A., 'The Clean Development Mechanism in China: Assessing the Tension between Development and Curbing Anthropogenic Climate Change' (2009) 27(1) *Virginia Environmental Law Journal* 91.

Schipper, Lisa, 'Conceptual History of Adaptation in the UNFCCC Process' (2006) 15(1) *Review of European, Comparative and International Environmental Law* 82.

Schreurs, Miranda, 'Multi-Level Climate Governance in China' (2017) 27 *Environmental Policy and Governance* 163.

Schuman, Sara, and Alvin Lin, 'China's Renewable Energy Law and Its Impact on Renewable Power in China: Progress, Challenges and Recommendations for Improving Implementation' (2012) 51 *Energy Policy* 89.

Shan, Yuli et al., 'China CO_2 Emission Accounts 1997–2015' (2018) 5 *Scientific Data* 1.

Shan, Yuli et al., 'New Provincial CO_2 Emission Inventories in China Based on Apparent Energy Consumption Data and Updated Emission Factors' (2016) 184 *Applied Energy* 742, 743.

Shao, Daoping, 'Foreign Legislation on Climate Change and Its Inspiration for China' (国外应对气候变化立法及对中国的启示) (2010) 17 *Environmental Protection* (环境保护) 66.

Shi, Xueying, and Yang Bowen, 'Penalty Rules of Regulated Enterprises Failing to Achieve Performance Targets' (控排企业碳交易未达履约目标的罚则设定) (2018) 28(4) *China's Population, Resources and Environment* (中国人口·资源与环境) 35.

Sijm, Jos, 'The Performance of Feed-in Tariffs to Promote Renewable Electricity in European Countries' (November 2002, ECN Project on Renewable Electricity Trends in European Countries) <www.ecn.nl/docs/library/report/2002/c02083.pdf?origin=publication_deta>.

Slater, Huw et al., *2018 China Carbon Pricing Survey* (China Carbon Forum, 2018) <www.eu-chinaets.org/upload/file/20180906/1536164718942317.pdf>.

Smil, Vaclav, *China's Environmental Crisis: An Enquiry into the Limits of National Development* (Routledge, 1993).

Song, Xixiang, and Gao Dali, 'Analysis of UK's Climate Change Act and Its Relevance to China' (论英国《气候变化法》及其对我国的启示) (2011) 2 *Journal of Shanghai University (Social Science Edition)* (上海大学学报(社会科学版)) 87.

Sprinz, Detlef, and Tapani Vaahtoranta, 'The Interest-Based Explanation of International Environmental Policy' (1994) 48(1) *International Organization* 77.

Stanway, David, 'China Blocks New Solar in 3 NW Regions Amid Overcapacity Fears' (15/02/2015, Reuters) <www.reuters.com/article/us-china-solarpower/china-blocks-new-solar-in-3-nw-regions-amid-overcapacity-fears-idUSKCN1Q404G>.

Stern, Nicholas H., 'Raising Consumption, Maintaining Growth and Reducing Emissions: The Objectives and Challenges of China's Radical Change in Strategy and its Implications for the World Economy' (2011) 12(4) *World Economics* 13.

Stoll, Peter-Tobias, 'The Climate as a Global Commons', in Daniel A. Farber and Marjan Peeters (eds), *Climate Change Law,* 131–141 (Elgar Encyclopedia of Environmental Law, Edward Elgar, 2016).

Su, Jack, Simone Hui, and Kevin Tsen, 'China Rationalizes Its Renewable Energy Policy' (2010) 23(3) *Electricity Journal* 26.

Suarez, Irene, and Jennifer Huang, 'Addressing Adaptation in a 2015 Climate Agreement' (Center for Climate and Energy Solutions, 2015).

Sun, Ying, 'The SPC Issued the Judicial Interpretation on Environmental Tort' (最高人民法院发布环境侵权司法解释) (China National Radio, 1 June 2015) <http://china.cnr.cn/gdgg/20150601/t20150601_518706638_1.shtml>.

Swart, Rob, and Frank Raes, 'Making Integration of Adaptation and Mitigation Work: Mainstreaming into Sustainable Development Policies?' (2007) 7(4) *Climate Policy* 288.

Tang, Ningning et al., 'Solar Energy Curtailment in China: Status Quo, Reasons and Solutions' (2018) 97 *Renewable and Sustainable Energy Reviews* 509.

Tang, Renhu et al., 'Key Challenges for the Establishment of the Monitoring, Reporting and Verification (MRV) System in China's National Carbon Emissions Trading Market' (2018) 18(S1) *Climate Policy* S106.

Tang, Shuang'e, 'GHG as An Air Pollutant: Debate in the US and Its Implications for China' (美国关于温室气体为"空气污染物"的争论及对我国的启示) (2011) 4 *Journal of Environmental Management College of China* (中国环境管理干部学院学报) 2.

Tangen, Kristian, Gørild Heggelund, and Jørund Buen, 'China's Climate Change Positions: At a Turning Point?' (2001) 12(2&3) *Energy & Environment* 237.

Teebken, Julie, 'Ten Years, Nine Floods: Local-Level Climate Adaptation in China' (News Security Beat, 29 March 2019) <www.newsecuritybeat.org/2018/03/ten-years-floods-local-level-climate-adaptation-china/>.

Thomas, Sebastian, Paul Dargusch, and Andrew Griffiths, 'The Drivers and Outcomes of the Clean Development Mechanism in China' (2011) 21 *Environmental Policy and Governance* 223.

Tian, Danyu, 'An Analysis of China's Current Climate Governance Organizations' (中国目前气候治理组织机构评析) (2013) 1 *Journal of China University of Political Science and Law* (中国政法大学学报) 147.

Tian, Danyu, 'Promote Legislation on Combating Climate Change, Form a Synergy of Systematic Legal Systems' (推动应对气候变化立法 形成系统法律制度合力) (China Environment News, 25 October 2019, 03) <http://epaper.cenews.com.cn/html/1/2019-10/25/03B/2019102503B_pdf.pdf>.

Tian, Danyu, 'Research on Climate Change Legislation' (应对气候变化立法研究) (2018) 3 *World Environment* (世界环境) 59.

Tian, Danyu, 'Studies on Local Laws on Climate Change' (中国地方应对气候变化立法研究) (2018) 3 *Law-Based Society* (法治社会) 80.

Tian, Danyu, 'Study of Legislation on Climate Change' (应对气候变化立法研究) (2018) 3 *World Environment* (世界环境) 62.

Tosun, Jale, 'Environmental Monitoring and Enforcement in Europe: A Review of Empirical Research' (2012) 22 *Environmental Policy and Governance* 437.

Trombetta, Maria Julia, 'Securitization of Climate Change in China: Implications for Global Climate Governance' (2019) 5 *China Quarterly of International Strategic Studies* 106.

Turnhout, Esther, Katja Neves, and Elisa de Lijster, '"Measurementality" in Biodiversity Governance: Knowledge, Transparency, and the Intergovernmental Science-Policy Platform on Biodiversity and Ecosystem Services (IPBES)' (2014) 46 *Environment and Planning A* 581.

Turnhout, Esther, Margaret M. Skutsch, and Jessica de Koning, 'Carbon Accounting', in Karin Bäckstrand and Eva Lövbrand (eds), *Research Handbook on Climate Governance,* 366–376 (Edward Elgar, 2015).

UNEP, 'Policies that Tackle Climate and Air Pollution at the Same Time Can Raise Global Climate Ambition' (UNEP, 18 October 2019) <www.unenvironment.org/news-and-stories/story/policies-tackle-climate-and-air-pollution-same-time-can-raise-global-climate>.

UNEP, 'The Status of Climate Change Litigation: A Global Review' (UNON Publishing Services Section, 2017).

Urban, Frauke, 'China's Rise: Challenging the North-South Technology Transfer Paradigm for Climate Change Mitigation and Low Carbon Energy' (2018) 113 *Energy Policy* 320.

Verschuuren, Jonathan, 'Legal Aspects of Climate Change Adaptation', in Erkki Hollo, Kati Kulovesi, and Michael Mehling (eds), *Climate Change and the Law,* 257–285 (Springer, 2012).

Vincent, Katharine, 'Uncertainty in Adaptive Capacity and the Importance of Scale' (2007) 17(1) *Global Environmental Change* 12.

Viñuales, Jorge, 'Legal Techniques for Dealing with Scientific Uncertainty in Environmental Law' (2014) 43 *Vanderbilt Journal of Transnational Law* 437.

Wang, Alex, 'Climate Change Policy and Law in China', in Cinnamon P. Carlarne, Kevin R. Gray, and Richard Tarasofsky (eds), *Oxford Handbook of International Climate Change Law,* 635–669 (Oxford University Press, 2016).

Wang, Alex, 'On the Evolution of Chinese Environmental Law and Governance' (2019) Unpublished Conference Paper available at SSRN <https://ssrn.com/abstract=3469184>.

Wang, Alex, 'The Search for Sustainable Legitimacy: Environmental Law and Bureaucracy in China' (2013) 37(2) *Harvard Environmental Law Review* 367.

Wang, Binhui, 'The Development of Carbon Emission Trading and Its Legislative Follow-up' (我国碳排放权交易的发展及其立法跟进) (2015) 13(2) *Current Law Science* (时代法学) 13.

Wang, Canfa, and Chen Yijian, '"Climate Justice" and China's Climate Change Legislation: Legislative Goal and Institutional Choice' ("气候正义"与中国气候变化立法的目标和制度选择) (2014) 2 *Social Science and Chinese Universities* (中国高校社会科学) 125.

Wang, Canfa, and Liu Zhe, 'The Legislation Model of Responding to Climate Change in China' (论我国应对气候变化立法模式的选择) (2015) 6 *Journal of China University of Political Science and Law* (中国政法大学学报) 113.

Wang, Guofei, 'Obstacle and Breakthrough: Reflections on the Carbon Tax Included in the Environmental Protection Tax' (障碍与突围:碳税纳入环境保护税的思考) (2015) 43 *Environmental Protection* (环境保护) 37.

Wang, Hanjie and others, 'Wind Power in China: A Cautionary Tale' (International Institute for Sustainable Development, 2016) <www.iisd.org/sites/default/files/publications/wind-power-in-china-cautionary-tale.pdf>.

Wang, Hui, 'Carbon Emission Right as Concession' (论碳排放权的特许权本质) (2017) 6 *Law and Social Development* (法制与社会发展) 171.

Wang, Hui, 'Local Level Climate Legislation in the US and Its Implications for China' (美国地方气候变化立法及其启示) (2017) 17(1) *The Journal of China University of Geosciences (Social Science Edition)* (中国地质大学学报(社会科学版)) 56.

Wang, Jian, 'Thoughts on Legal Problems in the Global Response to Climate Change' (全球气候变化背景下我国法律应对问题的思考) (2016) 2 *Theory Research* (学理论) 110.

Wang, Kai, 'Discussion on the Recording of Normative Documents' (论规范性文件的备案审查) (2010) 2010 *Zhejiang Social Sciences* (浙江社会科学) 11.

Wang, Mingyuan, 'Carbon Emission Right as A Quasi-Property Right and Developmental Right' (论碳排放权的准物权和发展权属性) (2010) 6 *China Legal Science* (中国法学) 92.

Wang, Mingyuan, 'Climate Change: An Environmental Problem or a Developmental Problem' (气候变化,是环境问题还是发展问题?) (2016) Z1 *Environmental Economics* (环境经济) 42.

Wang, Mingyuan, and Wang Shekun, 'Making a Specialized Climate Change Law' (制定专门的气候变化基本法) *Chinese Social Science Newspaper* (8 December 2011).

Wang, Pu, Lei Liu, and Tong Wu, 'A Review of China's Climate Governance: State, Market and Civil Society' (2018) 18 *Climate Policy* 664.

Wang, Qiang, 'Effective Policies for Renewable Energy – Example of China's Wind Power – Lessons for China's Photovoltaic Power' (2010) 14 *Renewable and Sustainable Energy Reviews* 702.

Wang, Shu, and Maosheng Duan, 'China' Domestic Offset Scheme' (Partnership for Market Readiness, World Bank, 2 March 2014) <www.thepmr.org/system/files/documents/China%27s%20domestic%20offset%20scheme%20-20140226_0.pdf>.

Wang, Xi, *Environmental Law in China* (Kluwer Law International, 2012).

Wang, Xi, and Tang Tang, 'Research and Scholarship on Climate Change Law in Developing Countries', in Daniel A. Farber and Marjan Peeters (eds), *Climate Change Law*, 115–125 (Elgar Encyclopedia of Environmental Law, Edward Elgar, 2016).

Wang, Xiahao, 'Study on Statute as a Branch of Legal Resource System' (论作为法的渊源的制定法) (2017) 35 *Tribune of Political Science and Law* (政法论坛) 51.

Wang, Yan et al., 'Environmental Impact Assessment of Projects in the People's Republic of China: New Law, Old Problems' (2003) 23 *Environmental Impact Assessment Review* 543.

Wang, Youling, and Lei Min, 'NDRC: China Has Basically Achieved Its Energy-Saving Targets During the Eleventh Five-Year Plan Period' (国家发改委：中国基本实现"十一五"规划节能降耗目标) 14 January 2011 <www.chinaenergycouncil.org/policynews/2011/1/14/2011114134037.html>.

Wang, Zhongying et al., 'The Retrospective Review and Assessment of China's Renewable Energy Law Implementation' (NDRC Energy Research Institute and Beijing Oriental Environmental Research Institute, 2006).

Watts, Jonathan, 'China Resorts to Blackouts in Pursuit of Energy Efficiency', *The Guardian* (London, 20 September 2010) <www.theguardian.com/world/2010/sep/19/china-blackoutsenergy-efficiency>.

Wen, Hui, 'Study on the Recording and Review System of Governmental Normative Documents' (政府规范性文件备案审查制度研究) (2015) 36 *Law Science Magazine* (法学杂志) 10.

Wu, Jing, and Zhang Yixin, 'Discussion on Integrating Climate Change Factors in Environmental Impact Assessment' (关于中国将气候变化因素融入环境影响评价的探讨) (2011) 9 *Environmental Pollution and Control* (环境污染与防治) 91.

Wu, Zongxin, and Zhihong Wei, 'Policies to Promote Energy Conservation in China' (1991) 19(10) *Energy Policy* 934.

Xia, Jun, Liu Chunzhen, and Ren Guoyu, 'Opportunity and Challenges of Climate Change Impacts on Water Resources' (气候变化对我国水资源影响研究面临的机遇与挑战) (2011) 26(1) *Advances in Earth Research* (地球科学进展) 1.

Xiao, Jinming, 'Restructuring the Relationship of Law and Policy for the Purpose of a Comprehensive Rule of Law' (为全面法治重构政策与法律关系) (2013) 5 *Chinese Public Administration* (中国行政管理) 38.

Xie, Laihui, 'Carbon Emission Trading or Carbon Tax: Theory and Policy' (碳交易还是碳税? 理论与政策) (2011) 6 *Financial Review* (金融评论) 103.

Xin, Chunpeng, 'The Socialist Legal System with Chinese Characteristics and Its Significance' (中国特色社会主义法律体系及其重大意义) (2014) 36 *Chinese Journal of Law* (法学研究) 19.

Xing, Huiqiang, 'Policy Growth and Law Hollowing: A Case Study of Economic Law' (政策增长与法律空洞化—以经济法为例的观察) (2012) 3 *Law and Social Development* (法制与社会发展) 120.

Xinhua News, 'Fading Subsidies, Rising Sustainability in China's Solar Energy Industry' (23 July 2019) <http://global.chinadaily.com.cn/a/201907/23/WS5d36c13ca310d830564008fb.html>.

Xinhua News, 'New Policies Support Ailing Solar Industry' (29 October 2012) <www.chinadaily.com.cn/business/2012-10/29/content_15853949.htm>.

Xu, Nan, and Zhang Chun, 'What the World is Getting Wrong about China and Climate Change' (China Dialogue, 18 February 2013) <www.chinadialogue.net/article/show/single/en/5711-What-the-world-is-getting-wrong-about-China-and-climate-change>.

Xu, Yixiang, and Feng Cuihua, 'Initial Allocation of Carbon Emission Right: Legal Institutions and Content' (碳排放权初始分配法律制度的基本构造) (2013) 7 *Tribune of Study* (学习论坛) 70.

Xu, Yugao, 'Different Perspectives between the North and the South on CDM: Viewpoints, Modalities, and Implications' (Development Research Academy for the 21st Century, Tsinghua University, Beijing, 1999).

Yamin, Farhana and Joanna Depledge, *The International Climate Change Regime: A Guide to Rules, Institutions and Procedures* (Cambridge University Press, 2004).

Yamineva, Yulia, and Zhe Liu, 'Cleaning the Air, Protecting the Climate: Policy, Legal and Institutional Nexus to Reduce Black Carbon Emissions in China' (2019) 95 *Environmental Science & Policy* 1.

Yan, Lu, and Du Yanfei, 'China Will Not Establish a Separate Carbon Tax', *People Net* (21 March 2016) <http://energy.people.com.cn/n1/2016/0321/c71661-28214363.html>.

Yang, Fan, Zhang Ting, and Zhang Hao, 'Adjudicating Environmental Tort Cases in China: Burden of Proof, Causation, and Insights from 513 Court Decisions' (2018) 21 *Asia Pacific Journal of Environmental Law* 171.

Yang, Kaiqi, 'EPIL Filed by Procuratorates: Nearly 60,000 Cases a Year; Confronts the Administration with Toughness but Does not Break up the Relationship' (检察院环境公益诉讼：一年近六万件，硬碰硬但不撕破脸) *Southern Weekly* (22 March 2019) <www.infzm.com/content/146284>.

Yang, Zewei, 'Carbon Emission Right: A New Developmental Right' (碳排放权：一种新的发展权) (2011) 41(3) *Journal of Zhejiang University (Humanities and Social Science)* (浙江大学学报(人文社会科学版)) 40.

Yang, Zewei, 'Provincial Level Climate Change Legislation in the Post-Kyoto Era: A Case Study of Hubei Province' ("后京都时代" 中国省级应对气候变化立法研究 – 以湖北省为例) (2013) 15(5) *Journal of Jiangsu University (Social Science Edition)* (江苏大学学报(社会科学版)) 21.

Yao, Ying, 'The Approach of Regulating GHGs through Law on Prevention and Control of Atmospheric Pollution' (《大气污染防治法》规制温室气体排放的进路探析) (2014) 18 *Environmental Protection* (环境保护) 55.

Yazykova, Sofia, and Carl Bruch, 'Incorporating Climate Change Adaptation Into Framework Environmental Laws' (2018) 48 *Environmental Law Reporter* 10334.

Ye, Qi, Jiaqi Lu, and Mengye Zhu, 'Wind Curtailment in China and Lessons from the United States' (Brookings-Tsinghua Center for Public Policy, 2018) <www.brookings.edu/wp-content/uploads/2018/03/wind-curtailment-in-china-and-lessons-from-the-united-states.pdf>.

Ye, Qi, and Tong Wu, 'China's "Yes" to New Role in Climate Battle' (Brookings, 4 December 2015) <www.brookings.edu/articles/chinas-yes-to-new-role-in-climate-battle/>.

Ye, Qi, and Tong Wu, 'The Politics of Climate Change in China' (2003) 4 *Wiley Interdisciplinary Reviews: Climate Change* 303.

Yin, Daphne, 'China's Carbon Emissions Traders Await Offset Demand' (Ecosystem Marketplace, 9 July 2013) <www.ecosystemmarketplace.com/pages/dynamic/article.page.php?page_id=9820>.

Yu, Hongyuan, 'The Paris Climate Agreement and China's Role in Global Climate Governance' (2016) 2 *China Quarterly of International Strategic Studies* 185.

Yu, Wenhao, 'Discussions on Duties of the Constitution and Law Committee on Constitutional Review' (宪法和法律委员会合宪性审查职责的展开) (2018) *China Legal Science* (中国法学) 61.

Yu, Wenxuan, and Tian Danyu, 'Climate Change Legislation in the US and Mexico and Its Implications for China' (美国和墨西哥应对气候变化立法及其借鉴意义) (2016) 18(2) *Journal of Jiangsu University (Social Science Edition)* (江苏大学学报(社会科学版)) 1.

Zahar, Alexander, 'Implementation of the Polluter Pays Principle in China' (2018) 27(3) *Review of European, Comparative and International Environmental Law* 293.

Zahar, Alexander, *International Climate Change Law and State Compliance* (Routledge, 2015).

Zeng, Ming, Chen Li, and Lisha Zhou, 'Progress and Prospective on Policy System of Renewable Energy in China' (2013) 20 *Renewable and Sustainable Energy Reviews* 36.

Zeng, Ming et al., 'Overall Review of Renewable Energy Tariff Policy in China: Evolution, Implementation, Problems and Countermeasures' (2013) 25 *Renewable and Sustainable Energy Reviews* 260.

Zeng, Xuelan et al., 'Data-Related Challenges and Solutions in Building China's National Carbon Emissions Trading Scheme' (2018) 18(S1) *Climate Policy* S90.

Zhai, Qing, 'Pushing the Modernization of China's Governance System and Capacity of Environmental Governance to Win the Tough Fight against Environmental Pollution' (推进生态环境治理体系和治理能力现代化 为打好污染防治攻坚战提供坚强保障) (2018) 10 *China Institutional Reform and Management* (中国机构改革与管理) 14.

Zhang, Da et al., 'Emissions Trading in China: Progress and Prospects' (2014) 75 *Energy Policy* 9.

Zhang, Guizhi, 'Results Anounced of the Eighth "Top Ten Public Interest Lawsuits in China"' (第八届" 中国十大公益诉讼" 评选结果出炉) (CBCGDF, 22 April 2019) <www.cbcgdf.org/NewsShow/4856/8369.html>.

Zhang, Haibin, 'China and International Climate Change Negotiations' (March 2013) WeltTrends Online-Dossier <http://welttrends.de/res/uploads/Zhang_China-and-International-climate-change-negotiations.pdf>.

Zhang, Hao, 'Antinomic Policy Making under Fragmented Authoritarianism: Regulating China's Electricity Sector through the Energy-Climate-Environment Dimension' (2019) 128 *Energy Policy* 162.

Zhang, Hao, 'China's Energy Conservation and Carbon Emissions Reduction System: Development and Status Quo of the Regulatory and Institutional Framework' (2012) 42(3) *Environmental Law Reporter* 10260.

Zhang, Hao, 'Prioritizing Access of Renewable Energy to the Grid in China: Regulatory Mechanisms and Challenges for Implementation' (2019) 3(2) *Chinese Journal of Environmental Law* 167.

Zhang, Hao, and Christopher Arup, 'Beyond the CDM: Regulating China's Domestic Offset Scheme' (2015) 45(1) *Environmental Law Reporter* 10049.

Zhang, Hao, and Ping Xu, 'Designing Regulation for China's Emission-Trading Pilot Programs through Trial and Error: An Effective Approach?' (2017) 7(2–3) *Climate Law* 125.

Zhang, Hong, 'State Policies as a Source of Civil Law' (2017) 38 *Social Sciences in China* 67.

Zhang, Jian, 'China's Energy Security: Prospects, Challenges, and Opportunities' (July 2011, Brookings) <www.brookings.edu/wp-content/uploads/2016/06/07_china_energy_zhang_paper.pdf>.

Zhang, Jun, 'Explore PIL According to the Law; Fulfil the Lofty Mission of Protecting Public Interests' (张军： 依法开拓公益诉讼，履行公益保护崇高使命) (Xinhua Net, 12 March 2019) <www.xinhuanet.com/politics/2019lh/2019-03/12/c_1210079657.htm>.

Zhang, Lei, 'GHG Emission Right as Property Right and Its Institutional Dilemma: Refection of Hardin's "The Tragedy of Commons"' (温室气体排放权的财产权属性和制度化困境 – 对哈丁" 公地悲剧"理论的反思) (2011) 1 *Law and Social Development* (法制与社会发展) 101.

Zhang, Mo, 'The Socialist Legal System with Chinese Characteristics: China's Discourse for the Rule of Law and a Bitter Experience' (2010) 24 *Temple International & Comparative Law Journal* 1.

Zhang, Qianfan, 'A Constitution without Constitutionalism? The Paths of Constitutional Development in China' (2010) 8 *International Journal of Constitutional Law* 950.

Zhang, Qijing, 'Research on China's Legislation to Address Climate Change in the Post-Paris Era' (后巴黎时代中国应对气候变化立法研究) (2017) 29 *Environment and Development* (环境与发展) 2.

Zhang, Qing, *Ground-breaking but not so Effective: Environmental Public Interest Litigation Filed by NGOs in China* (PhD Thesis, The Chinese University of Hong Kong, 2019).

Zhang, Sufang et al., 'Interactions between Renewable Energy Policy and Renewable Energy Industrial Policy: A Critical Analysis of China's Policy Approach to Renewable Energy' (2013) 62 *Energy Policy* 342.

Zhang, Sufang et al., 'The Development Trajectories of Wind Power and Solar PV Power in China: A Comparison and Policy Recommendations' (2013) 26 *Renewable and Sustainable Energy Reviews* 322.

Zhang, Sufang, Phillip Andrews-Speed, and Meiyun Ji, 'The Erratic Path of the Low-Carbon Transition in China: Evolution of Solar PV Industry' (2014) 67 *Energy Policy* 903.

Zhang, Sufang, Phillip Andrews-Speed, and Sitao Li, 'To What Extent Will China's Ongoing Electricity Market Reforms Assist the Integration of Renewable Energy?' (2018) 114 *Energy Policy* 165.

Zhang, Sufang, and Yongxiu He, 'Analysis on the Development and Policy of Solar PV Power in China' (2013) 21 *Renewable and Sustainable Energy Reviews* 393.

Zhang, Zhijun et al., 'Costs of Carbon Sequestration for Afforestation/Reforestation Projects Under the Clean Development Mechanism: A Case Study of Pearl River Basin in Guangxi, China (清洁发展机制（CDM）造林再造林项目碳汇成本研究—以CDM广西珠江流域治理再造林项目为例)' (2009) 5 *Climate Change Research* (气候变化研究进展) 348.

Zhang, Zhongli, 'Carbon Trading Law in South Korea and Its Application to China' (韩国碳排放交易法律及其对我国的启示) (2016) 5 *Northeast Asia Forum* (东北亚论坛) 50.

Zhang, Zhongli, 'Legal System Construction of GHG Emission Control' (论温室气体排放控制法律制度体系的建构) (2016) 2 *Tsinghua Forum of Rule of Law* (清华法制论衡) 382.

Zhang, Zhongxiang, 'Carbon Emissions Trading in China: The Evolution from Pilots to a Nationwide Scheme' (2015) 15 *Climate Policy* 104.

Zhang, Zhongxiang, 'China in the Transition to A Low-Carbon Economy' (2010) 38 *Energy Policy* 6638.

Zhang, Zhongxiang, 'Climate Mitigation Policy in China' (2015) 15(S1) *Climate Policy* S1.

Zhang, Zitai, 'Preliminary Study of China's Legal System on Climate Change' (中国气候变化应对法框架体系初探) (2010) 5 *Journal of Nanjing University* (南京大学学报) 37.

Zhao, Jun, 'Legal Principles of Climate Change Legislation' (我国应对气候变化立法的基本原则研究) (2015) 7 *Politics and Law* (政治与法律) 80.

Zhao, Jun, and Sun Songlong, 'An Assessment of Incorporating GHGs within Law on Prevention and Control of Atmospheric Pollution' (《大气污染防治法》将温室气体排放纳入控制体系的评析) (2016) 33(1) *Journal of Shanghai University (Social Science Edition)* (上海大学学报(社会科学版)) 27.

Zhao, Yue, Lyu Shuang, and Wang Zhu, 'Prospects for Climate Change Litigation in China' (2019) 8 *Transnational Environmental Law* 349.

Zhao, Yuhong, 'Assessing the Environmental Impact of Projects: A Critique of the EIA Legal Regime in China' (2009) 49 *Natural Resources Journal* 493.

Zhao, Yuhong, 'Public Participation in China's EIA Regime: Rhetoric or Reality?' (2010) 22(1) *Journal of Environmental Law* 89.

Zhao, Zhenyu, Shuang-ying Zhang, and Jian Zuo, 'A Critical Analysis of the Photovoltaic Power Industry in China – From Diamond Model to Gear Model' (2011) 15(9) *Renewable and Sustainable Energy Reviews* 4963.

Zheng, Heran et al., 'How Modifications of China's Energy Data Affect Carbon Mitigation Targets' (2018) 116 *Energy Policy* 337.

Zheng, Jinggao, and Tian Ye, 'From "State Will" to "Rule of Law" – Extensive Viewpoints of Law on the Relationship between Law and Policy' (从 "国家意志" 到 "行政法治" – 在法律与政策关系上的泛法律观) (2007) 7 *Journal of China University of Geosciences (Social Sciences Edition)* (中国地质大学学报:社会科学版) 100.

Zheng, Shuang, and Dou Yong, 'Coping with Climate Change with Economic Instruments: Comparison between Carbon Tax and Carbon Trading' (利用经济手段应对气候变化 – 碳税与碳交易对比分析) (2013) 35(10) *China Energy* (中国能源) 11.

Zhong, Shuying, Liu Chi, and Qin Liqiong, 'Solar Industry Development and Policy Support in China' (2011) 5 *Energy Procedia* 768.

Zhong, Xiaojian et al., 'The Features, Motivations and Trends of China's Carbon Trading Program: A Comparative Study Based on International Experiences' (中国碳交易市场

的特征、动力机制与趋势 – 基于国际经验比较) (2017) 37 *Acta Ecologica Sinica* (生态学报) 331.

Zhou, Chen, *The Legal Barriers to Technology Transfer under the UN Framework Convention on Climate Change* (Springer, 2019).

Zhou, Xian, and Chen Demin, 'Rethinking and Reconstructing Public Participation in Climate Change Response' (公众参与气候变化应对的反思与重塑) (2019) 29 *China Population, Resources and Environment* (中国人口·资源与环境) 117.

Zhou, Xiaoguang, and Zhang Jianwei, 'Thoughts on the Coordinated Governance of Air Pollution and Climate Change' (关于大气污染与气候变化协同治理的法律思考) (2017) 5 *Tribune of Social Sciences* (社会科学论坛) 222.

Zhu, Deli, 'The Risks and Legal Response of the Carbon Trading Scheme' (碳排放权交易机制的风险挑战及其法律应对) (2018) 4 *Studies in Dialectics of Nature* (自然辩证法研究) 78.

Zhu, Liu et al., 'Reduced Carbon Emission Estimates from Fossil Fuel Combustion and Cement Production in China' (2015) 524 *Nature* 335.

Zhu, Liu et al., 'Steps to China's Carbon Peak' (2015) 522 *Nature* 279.

Zhu, Qian, 'The Legal Nature of Environmental Public Interests' (环境公共利益的法律属性) (2016) 2 *Study & Exploration* (学习与探索) 59.

Zhu, Yan, 'Asking again about Carbon Reduction, Will There Be Carbon Tax?' (再问碳减排，何日" 税"会来) *China Energy Newspaper* (22 October 2018).

Zimmerman, James M., *China Law Deskbook: A Legal Guide for Foreign-Invested Enterprises* (3rd edn, American Bar Association, 2010).

Zou, Hongyang et al., 'Market Dynamics, Innovation, and Transition in China's Solar Photovoltaic (PV) Industry: A Critical Review' (2017) 69(C) *Renewable and Sustainable Energy Reviews* 199.

Zou, Pingxue, 'Missions and Institutional Innovation of the Constitution and Law Committee' (宪法和法律委员会的目标定位与机制创新) (2018) 4 *China Law Review* (中国法律评论) 47.

Index

For Product Safety Concerns and Information please contact our EU representative GPSR@taylorandfrancis.com
Taylor & Francis Verlag GmbH, Kaufingerstraße 24, 80331 München, Germany

www.ingramcontent.com/pod-product-compliance
Ingram Content Group UK Ltd.
Pitfield, Milton Keynes, MK11 3LW, UK
UKHW021302190926
475701UK00009B/577
* 9 7 8 0 3 6 7 5 0 9 7 0 5 *